Learn Java with Math

Using Fun Projects and Games

Ron Dai

Apress®

Learn Java with Math: Using Fun Projects and Games

Ron Dai
Seattle, WA, USA

ISBN-13 (pbk): 978-1-4842-5208-6   ISBN-13 (electronic): 978-1-4842-5209-3
https://doi.org/10.1007/978-1-4842-5209-3

Copyright © 2019 by Ron Dai

Managing Director, Apress Media LLC: Welmoed Spahr
Acquisitions Editor: Steve Anglin
Development Editor: Matthew Moodie
Coordinating Editor: Mark Powers

Cover designed by eStudioCalamar

Cover image designed by Freepik (www.freepik.com)

Distributed to the book trade worldwide by Springer Science+Business Media New York, 233 Spring Street, 6th Floor, New York, NY 10013. Phone 1-800-SPRINGER, fax (201) 348-4505, e-mail orders-ny@springer-sbm.com, or visit www.springeronline.com. Apress Media, LLC is a California LLC and the sole member (owner) is Springer Science + Business Media Finance Inc (SSBM Finance Inc). SSBM Finance Inc is a **Delaware** corporation.

For information on translations, please e-mail editorial@apress.com; for reprint, paperback, or audio rights, please email bookpermissions@springernature.com.

Apress titles may be purchased in bulk for academic, corporate, or promotional use. eBook versions and licenses are also available for most titles. For more information, reference our Print and eBook Bulk Sales web page at http://www.apress.com/bulk-sales.

Any source code or other supplementary material referenced by the author in this book is available to readers on GitHub via the book's product page, located at www.apress.com/9781484252086. For more detailed information, please visit http://www.apress.com/source-code.

Printed on acid-free paper

To Angela, Henry, and Hanson,
who always brighten my world

Table of Contents

TABLE OF CONTENTS

About the Author

Ron Dai is a software engineer and data scientist at Microsoft. He is also a mathematics and computer science instructor at NWCS (Northwest Chinese School, http://www.nwchinese.org) located in Bellevue, Washington. He enjoys teaching computer science using math. He has written a book titled *Cool Math – Scenarios and Strategies,* which is available on amazon.com.

About the Technical Reviewer

 **Jeff Friesen** is a freelance teacher and software developer with an emphasis on Java. In addition to authoring *Java I/O, NIO and NIO.2* (Apress, 2015), *Java Threads and the Concurrency Utilities* (Apress, 2015), and the first edition of this book, Jeff has written numerous articles on Java and other technologies (such as Android) for JavaWorld (JavaWorld.com), informIT (InformIT. com), Java.net, SitePoint (SitePoint.com), and other web sites. Jeff can be contacted via his web site at JavaJeff.ca or via his LinkedIn (LinkedIn. com) profile (`www.linkedin.com/in/javajeff`).

Acknowledgments

I am grateful for my wife, Angela, who has always been supportive of my technical research and teaching work in so many ways. I am also thankful to my two lovely and intelligent sons, Henry and Hanson, who have provided me firsthand feedback and inspired me when I created the idea of this book. I can never forget about what I have learned from Mr. Bangfu Mo and Mr. Jim Pierson, who have provided me with guidance on how to elaborate the creative ideas in a book. I also want to take this opportunity to send my sincere gratitude for their invaluable advice that they have given me in the past. My special thanks go to everyone on the Apress editorial team, who are all world-class professionals.

(Designed by Hanson Dai)

Preface

Congratulations on finding out about this book. I am writing this book to help beginners learn Java programming effectively and with plenty of fun. The book is designed to simplify the complexity and guide the learner to explore and discover things under the hood. I hope the instructions inside this book are intuitive enough for beginners to follow through with hands-on practice.

Having a good foundation of math skills is undoubtedly super powerful when learning programming. In the meantime, it is a good opportunity to practice mathematical problem solving when you study programming. With this motivation, I have included some math practice problems applicable to programming-related concepts.

The more practice you do, the more effectively you will be able to produce results. It requires intensive and extensive problem solving exercises for anyone to master coding skills. In this book, I have included quite a few interesting coding problems for practice. I hope you will have an enjoyable learning experience. Source Code for this book is accessible via the Download Source Code button located at www.apress.com/9781484252086.

PART I

Java Basic

CHAPTER 1

Introduction

There are many good Java programming books on the market, but it is not easy to find one fit for a beginner who is new to Java and has minimal programming knowledge.

This book will help beginners learn how to effectively program in Java. My intent is to simplify the more complex aspects of Java and to guide the learner in exploring things "under the hood." I hope the instructions inside this book are intuitive enough for readers to follow through with hands-on practice.

People who have experience with programming understand that mathematical knowledge plays a crucial role in programming design. So having a good foundation of math skills is undoubtedly super helpful when learning programming. This book provides a good opportunity to practice mathematical problem solving in a programming context. With this motivation in mind, I have included some math practice problems applicable to programming-related concepts.

Learning with deliberate practice enhances your understanding of new concepts on a deeper level. Actively participating in hands-on projects is a critical part of the learning process. And more practice will lead to producing results more quickly. I hope this will be an enjoyable learning experience for you.

© Ron Dai 2019
R. Dai, *Learn Java with Math*, https://doi.org/10.1007/978-1-4842-5209-3_1

Programming work involves designing and writing code using a certain computer language. Correctly executed code will perform repetitive tasks and accomplish expected goals. Nowadays, as high-technology products are being integrated into our daily lives, computer programming skills are becoming indispensable almost everywhere. Many daily computation jobs have already been replaced by programmed devices–you don't need to look further than the self-checkout line at your local supermarket or the ever-increasing number of products purchased online. Reoccurring events are increasingly controlled by automated systems, such as building security system, thermostats mounted on the walls inside your house, and a plethora of other examples.

Another example is gaming software, which has such a rich user interface that many of us—from teenagers to adults—are already addicted to it. All these products and services are essentially built by computer programming.

As the beauty of artificial intelligence emerges, we can already see and feel the power of applications of computer technology more than ever before. If you have watched Hollywood movies like *Arrival* or *Passengers* (both released in 2016), I am sure you were fascinated by the amazingly intelligent robots depicted in the movies. If you are curious how a computer can precisely recognize an object with an activity in any picture, I suggest you listen to an exciting TED Talk named "How We Teach Computers to Understand Pictures" All of these amazing things are empowered by software, which is written in programming language(s).

To become a good programmer, you need to understand logical control and basic counting methods. It will require more sophisticated math knowledge if you want to develop a system to control objects' activities.

There are quite a lot of famous but unsolved problems in math history. As computer technology improves, we can leverage computers' talents to solve some of these problems.

For example, the Collatz conjecture states that if you randomly pick a positive integer N, and if it is even, divide it by 2; if it is odd, multiply it by 3 and add 1. And if you repeat this procedure long enough, eventually the end result of N will always be 1.

Mathematicians and data researchers have tried millions of numbers. No exception has been found, but no one has found a way to prove all integers following this rule.

Using simple Java programming, we can prove the Collatz conjecture for any positive integer up to N. In the following short program, I will test the conjecture with every integer and find out its sequence length, which is the number of operations for it to reach the result "1."

```java
public class ProveIt {
        public static void main(String[] args) {
                // representation of a million
                final long N = 1000 * 1000;
                for(long i = 1; i <= N; i++) {
                        System.out.println("i=" + i + " - " +
                                GetCollatzSequenceCount(i));
                }
                System.out.println("DONE!");
        }

        private static long GetCollatzSequenceCount(long n) {
                if (n <= 0) return 0;
                long count = 0;
                while(true) {
                        if (n == 1) return count;
                        if (n % 2 == 0) {
                                n /= 2;
```

```
                    } else {
                            n = n * 3 + 1;
                    }
                    count++;
            }
        }
}
```

Guess what? To test up to 1,000,000 integers, it completes executions and reports results back within several seconds on a normal work laptop. Don't worry about understanding or running this code now; just appreciate that this short program can churn through 1,000,000 iterations in only a few seconds.

The last part of the output is:

```
i=999991 - 165
i=999992 - 113
i=999993 - 165
i=999994 - 113
i=999995 - 258
i=999996 - 113
i=999997 - 113
i=999998 - 258
i=999999 - 258
i=1000000 - 152
DONE!
```

One last thing to mention about notation in this book:

> **Math**: describes a specific math concept.

> **Problems**: provides a list of follow-up exercises. You can find hints for some problems.

Hint: suggests ideas for references to solve the problem.

Finally, students are encouraged to try ***Lab Work***, after learning ***Answer*** and ***Example***.

Problems

1. List an example that you have observed about something satisfying both (a) and (b) described as below.

    (a) There is no programming feature associated with it now.

    (b) It will function much more efficiently if there is a program built in it.

2. How do we exchange different types of water between the two cups?

    You are not allowed to mix the water.

3. I am thinking about an integer between 1 and 100. You may ask me questions in order to identify the integer, but you are not allowed to ask questions like "what is this integer?"

    What is your strategy to ask the minimum number of questions in order to figure out the number?

4. There are 27 ping pong balls. All of them look identical and weigh the same, except that one of them is lighter. Using a balance scale, how do you quickly find the one that is not the same as the others?

5. How do you use the following four numbers with basic operators ("+", "-", "x", and "/") to create a math formula which equals 24? You may use each number only once, but you can use parentheses.

CHAPTER 2

Number Basics

What Is a Numeral System?

Many different numeral systems exist because there are specific uses where a certain numeral system is more convenient to use and offers advantages over others. For example:

- Weight: 1 pound = 16 ounces

- Length: 1 yard = 3 feet, 1 foot = 12 inches

- Babylonian numeral: Base 60

(from Wikipedia)

© Ron Dai 2019

R. Dai, *Learn Java with Math*, https://doi.org/10.1007/978-1-4842-5209-3_2

- In Ancient China: Ying/Yang – "binary," Ba Gua – 8 trigrams

(from Wikipedia)

- Decimal counting

  - Ten symbols: 0 – 9

- Binary counting

  - Two symbols: 0 and 1

- Time measurement

  One day = 24 hours

  One hour = 60 minutes = 3600 seconds

Why Do People Use Decimal Numbers, While Computers Use Binary Numbers?

A simple answer is that human beings have ten fingers and ten toes, but a computer has only two states.

Joking aside, a computer is built with many connections and components (parts) that are used to transfer and store data, as well as to communicate with other components. Most of the storing, transferring, and communicating events happen with digital electronics. Digital electronics use the binary system (ON or OFF). A signal with a series of ON/OFF pulses is equal to a binary number.

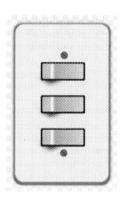

How to Convert a Number Between Different Numeral Systems

[Math] Conversion between Decimal and Binary:

(1) Convert a decimal number to a binary number

[Example]

Convert 350 in base 10 number, to a binary number (base 2)

[Answer]

In base 10, we can write 350 with this equation:

$$350 = 3 * 10^2 + 5 * 10^1 + 0 * 10^0$$

Notice each coefficient (i.e., 3, 5, and 0) is less than 10, and there is no coefficient for 10^3 or above.

Now we want to change it to something like this:

$$350 = a * 2^8 + b * 2^7 + c * 2^6 + d * 2^5 + e * 2^4 + f * 2^3 +$$
$$g * 2^2 + h * 2^1 + i * 2^0$$

Notice there is no 2^9 or above, because we know $350 < 512 = 2^9$

$350 - 1 * 256$ (i.e., 2^8) $= 94 < 128 = 2^7 \rightarrow a = 1, b = 0;$

$94 - 1 * 64$ (i.e., 2^6) $= 30 < 32 = 2^5 \rightarrow c = 1, d = 0;$

$30 - 1 * 16$ (i.e., 2^4) $= 14 \rightarrow e = 1;$

$14 - 1 * 8$ (i.e., 2^3) $= 6 \rightarrow f = 1;$

$6 - 1 * 4$ (i.e., 2^2) $= 2 \rightarrow g = 1;$

$2 - 1 * 2$ (i.e., 2^1) $= 0 \rightarrow h = 1, i = 0;$

Therefore, $350 = 1 * 2^8 + 0 * 2^7 + 1 * 2^6 + 0 * 2^5 + 1 * 2^4 +$
$$1 * 2^3 + 1 * 2^2 + 1 * 2^1 + 0 * 2^0$$

Which means $(350)_{10} = (101011110)_2$

The subscript number (10 and 2) indicates its number base.

(2) Convert a binary number to a decimal number

[Example]

Convert binary number 11001001 to a decimal number

[Answer]

We rewrite the expression of the binary number as shown below.

$(11001001)_2$

$$= 1 * 2^7 + 1 * 2^6 + 0 * 2^5 + 0 * 2^4 + 1 * 2^3 + 0 * 2^2 +$$
$$0 * 2^1 + 1 * 2^0$$

$$= 128 + 64 + 8 + 1$$

$$= (201)_{10}$$

To practice conversion between decimal and binary, I recommend this online game: `http://games.penjee.com/binary-numbers-game/`

[Math] Fractions in Decimal and Binary

(3) Convert a decimal point number (base 10) to a binary number

We need to understand how we identify each digit after the decimal point. For example, 4.3256

Remove integer part "4," so we have 0.3256.

0.3256 x 10 = 3.256 → 3 is the 1st digit after the decimal point

Remove integer part "3," so we now have 0.256

0.256 x 10 = 2.56 → 2 is the 2nd digit after the decimal point

Remove integer part "2," so we now have 0.56

0.56 x 10 = 5.6 → 5 is the 3rd digit after the decimal point

Remove integer part "5," so we now have 0.6

0.6 x 10 = 6 → 6 is the 4th digit after the decimal point

Remove integer part "6,"and we are done.

The same process applies when we convert a fraction from a decimal to a binary.

Integer part of "4.3256" is "4," which is 100 in binary.

From now on, we only look at the decimal part.

0.3256 x 2 = 0.6512 → 0 is the 1st digit after the decimal point

0.6512 x 2 = 1.3024 → 1 is the 2nd digit

0.3024 x 2 = 0.6048 → 0 is the 3rd digit

0.6048 x 2 = 1.2096 → 1 is the 4th digit

0.2096 x 2 = 0.4192 → 0 is the 5th digit

0.4192 x 2 = 0.8392 → 0 is the 6th digit

0.8392 x 2 = 1.6784 → 1 is the 7th digit

......

Repeat until we finally get 0, or we see a repeating pattern.

$(100.0101001...)_2$ is the final answer.

[Math] Binary arithmetic: Addition, Subtraction, Multiplication, Division, Square root

Binary addition and subtraction operations follow rules such as these:

0 + 0 = 0 → 0 - 0 = 0

0 + 1 = 1 → 1 - 0 = 1

1 + 0 = 1

1 + 1 = 0 (carry one) = 10 → 10 - 1 = 1

Note As opposed to the decimal numeral system (a.k.a. base 10 numbers) that we are familiar with, a binary number has 2 as its base and has only 0 or 1 as its representation for every digit. In an addition operation, when any digit reaches 2, it becomes "carry one" to its left digit. However, in a subtraction operation, a digit 0 will need to borrow 2 from its left digit to subtract 1. However, this is the opposite direction of the operation to the "carry one."

Binary multiplication and division operations follow rules as the following:

$0 \times 0 = 0$

$0 \times 1 = 0$

$1 \times 0 = 0$

$1 \times 1 = 1$

This is an example of division between binary numbers.

```
        11
11) 1011
    −11
    101
    −11
     10     →      remainder (r)
```

Conversion between binary and other numeral systems:

- Hexadecimal – base 16 number system

  Mapping between decimal and hexadecimal:

Hexadecimal:	0	1	2	3	4	5	6	7	8	9	A	B	C	D	E	F
Decimal:		0	1	2	3	4	5	6	7	8	9	10	11	12	13	14	15

Since every four digits in a binary forms one hexadecimal digit, to convert a binary number to its hexadecimal, we group every four digits in the binary from the right.

For example, 100 in binary equals 4 in hexadecimal and 1011 in binary is $8 + 2 + 1 = B$ in hexadecimal. Therefore, 1001011 in binary is 4B in hexadecimal.

- Octal – base 8 number system

Every three digits in a binary forms one octal digit. We can group every three digits in a binary from the very right and convert it to its octal form.

For example,

To convert 10111011 in binary to its octal result:

Step 1 - group it by every three digits from the right: $(10)_2(111)_2(011)_2$;

Step 2 - convert every group of up to three digits (0 to 1) to an octal digit (0 to 7): $(2)_8(7)_8(3)_8$;

Step 3 - the converted octal result is 273_8.

Inversely, to convert 273 in octal to its binary format, we convert each octal digit to a three-digit binary number:

$273_8 = (2)_8(7)_8(3)_8 = (010)_2(111)_2(011)_2 = 010111011_2$

What Is Bit, Byte, KB, MB, GB, TB, and PB?

Bit means a binary digit, 0 or 1. It is the smallest unit of data.

Byte is a sequence of eight bits.

1 Byte (= 8 Bit), KB, MB, GB, TB, PB

1,024 Bytes = 1 KB	KB:	Kilobyte
1,024 KB = 1 MB	MB:	Megabyte
1,024 MB = 1 GB	GB:	Gigabyte
1,024 GB = 1 TB	TB:	Terabyte
1,024 TB = 1 PB	PB:	Petabyte

What Is Bitwise?

In computers, an integer number is represented as a sequence of bits in memory. We usually interact with decimal numbers in display through a computer's graphic user interface. However, its binary forms carry out the actual calculations inside the computer. Bitwise is just a level of operations that involves working with individual bits.

Bitwise operators contain three basic ones:

& → AND

0 & 0 = 0 & 1 = 1 & 0 = 0

1 & 1 = 1

A logical AND (&) of each bit pair results in a 1, if the first bit is 1 AND the second bit is 1. Otherwise, the result is zero.

Examples:

01 & 00 = 00

11111111 & 01100101 = 01100101

| →   OR

    $0 \,|\, 0 = 0$

    $0 \,|\, 1 = 1 \,|\, 0 = 1 \,|\, 1 = 1$

    A logical OR (|) of each bit pair results in a 1,

(1)   if the first bit is 1 OR the second bit is 1.

(2)   or, if both the first and the second bit are 1.

    Otherwise, the result is zero.

    Examples:

    $0101 \,|\, 0011 = 0111$

    $0010 \,|\, 1000 = 1010$

^ →   NOT

    A unary operation performs a logical negation on each bit.

    In other words, after this operation, a 1 bit is flipped to a 0 bit and a 0 bit is flipped to a 1 bit.

    Examples:

    $\wedge\, 0011 = 1100$

    $\wedge\, 01010110 = 10101001$

Problems

1.   Why do computers use binary numbers?

2.   What do Hexadecimal, Octal, and Bitwise mean?

CHAPTER 3

Java Basics

Today, computers use so many different kinds of programming languages. Each of them plays an essential role in a certain area to solve a specific type of problem. Java was invented and developed by Sun Microsystems in the early 1990s; it was then acquired by Oracle, Inc. After 20 years of development, Java has now become one of the most popular programming languages in the world.

Java is a typical OOP, a.k.a. object-oriented programming language, which deals with a bunch of "objects." These objects contain data and operations. Operations are about what can be done to data within the object.

In the Java world, there is an open source development tool called Eclipse. It has rich features, and it is free to use. Eclipse is also a suitable tool for beginners. We will be using Eclipse to start exploration of the Java world.

What Features Does Java Have?

Object-Oriented

An object is a "thing" that has some attributes (a.k.a. Properties). The object performs a set of operations (a.k.a. Methods). The operations define behaviors of the object.

© Ron Dai 2019
R. Dai, *Learn Java with Math*, https://doi.org/10.1007/978-1-4842-5209-3_3

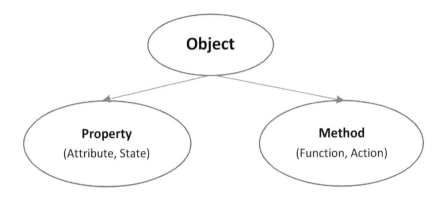

Class-Based

A *class* is simply a representation of a type of *object*. It is a template that describes details of an object. A class is composed of three elements: a name, attributes, and operations.

Java Bytecode

Java bytecode is the machine language of the Java virtual machine (JVM). Java bytecode will be translated to a machine-specific native code by the JVM, when it runs on that machine. So, on a Windows computer, the JVM bytecode is translated into Windows-specific native code; and on a Linux computer, it is translated into Linux-specific native code. This is called write once, run anywhere (shown in Figure 3-1).

Figure 3-1. *Write once, run anywhere*

When a JVM loads a class file, it gets one stream of bytecodes for each method in the class. The bytecodes for a method are executed when that method is invoked during the execution of the program.

Although it might be overwhelming to beginners, we also want to introduce some of the other powerful features built into the Java language as well as its runtime engine.

- Multi-threading

  Java language supports multi-threading capabilities, which enable multiple tasks running at the same time. This feature boosts up the computing power of the Java code and makes Java applications highly responsive.

- Secure code

  Java doesn't use pointers like in other languages (i.e., C or C++). This has avoided a traditional security loophole. In addition to runtime checking, Java does static-type checking using strict rules during compilation. Java has its exception handling to catch unexpected errors. Java provides a cryptographic security mechanism when users are getting code across networks and so on. These security functionalities make Java a more secure programming language than other ones.

- Garbage collection

  Java has its own uniquely designed memory-management mechanism. Unlike C/C++ language, Java doesn't require developers to take care of

memory management in terms of when to register
or free the memory. It will automatically collect
and free up the unused memories. This has made
development work much easier.

The last but not least powerful feature to mention is Java's super-rich
open source libraries. This is one of the big reasons why Java is increasingly
so popular among developers. And, because Java's developer community
is getting bigger and stronger, Java will evolve to be an even more powerful
programming language over time.

CHAPTER 4

Start Playing with Java

Download and install a Java runtime environment (choose the right version based on your computer's operating system): `https://www.java.com/en/download/manual.jsp`

Download and install Eclipse (look for the most recent stable version): `http://www.eclipse.org/downloads/`

After installation, you should see an icon as shown—the "Neon" version—as an example on your desktop.

Once you launch Eclipse, you will need to specify the Workspace (Figure 4-1).

© Ron Dai 2019
R. Dai, *Learn Java with Math*, https://doi.org/10.1007/978-1-4842-5209-3_4

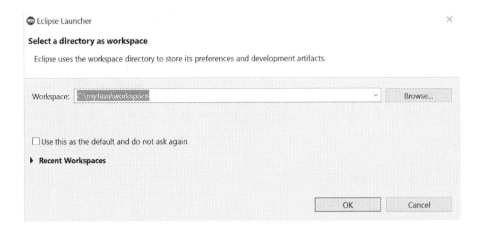

Figure 4-1. *Specifying the workspace*

What Is the Difference Between the JRE and the JDK?

The JRE is the "Java Runtime Environment." It is where your Java programs run. The JDK is the "Java Development Kit," which is the full-featured software development kit for Java, including JRE, the compiler, and tools (e.g., JavaDoc, Java debugger) to create and compile programs.

When you only want to run Java programs on your browser or computer, you will install the JRE. But if you want to do some Java programming, you will also need to install the JDK.

Figure 4-2 shows the clear relationship between the JRE and JDK, as well as their basic feature areas.

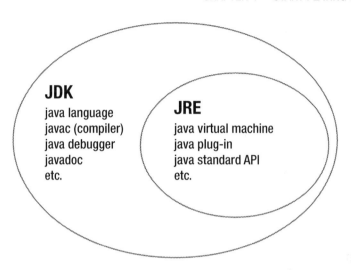

Figure 4-2. *JDK and JRE compared*

What Are a Workspace, Source, and Package?

"Workspace" is used to group a set of related projects together. Usually these projects will make up an application.

"Source" means source code, that is, the Java program and related code.

"Package" indicates a collection of files.

What Are Edit, Compile, and Execute?

"Edit" writes code in a Java language.

"Compile" converts Java source code to Java bytecode.

"Execute" runs the program.

- Edit → create "*.java" file
- Compile → generate "*.class" file

Creating Your First Program

Let's get started:

1. Once Eclipse is launched, left-click on "File" in
    the top menu bar and left-click on "New" in the
    drop-down menu. Then select "Java Project" from
    another drop-down menu as shown in Figure 4-3.

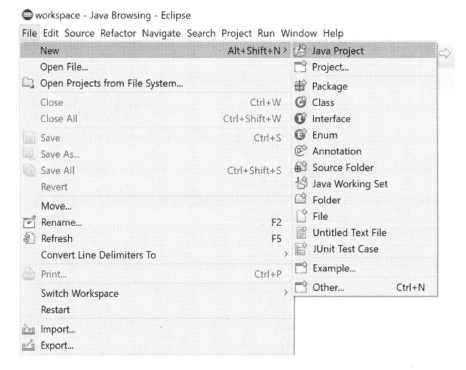

Figure 4-3. *Before we create a Java project or a Java class*

2. Now create a Java project named MyFirstProgram,
    as shown in Figure 4-4. Click "Finish."

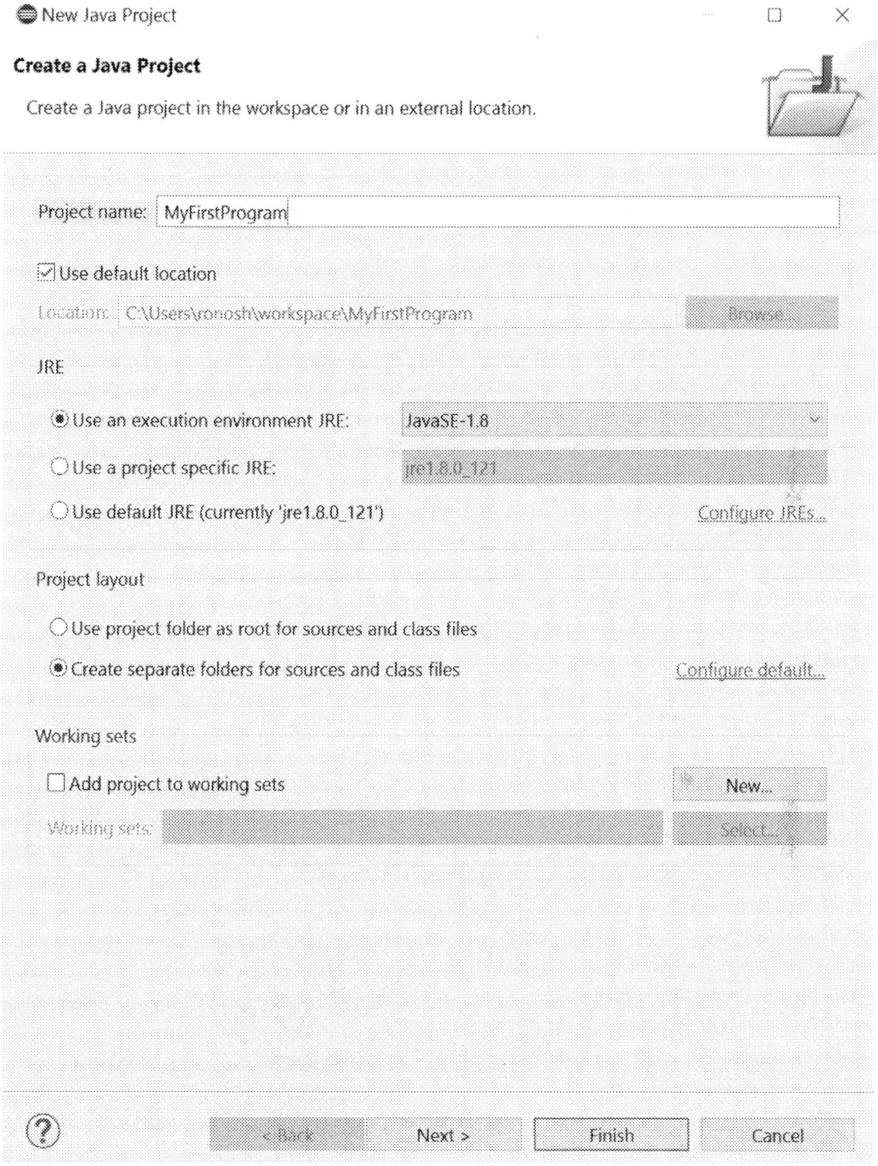

New Java Project   ☐   ✕

Create a Java Project

Create a Java project in the workspace or in an external location.

Project name: MyFirstProgram

☑ Use default location

Location: C:\Users\ronosh\workspace\MyFirstProgram   Browse...

JRE

⦿ Use an execution environment JRE: JavaSE-1.8   ⌄

◯ Use a project specific JRE: jre1.8.0_121

◯ Use default JRE (currently 'jre1.8.0_121')   Configure JREs...

Project layout

◯ Use project folder as root for sources and class files

⦿ Create separate folders for sources and class files   Configure default...

Working sets

☐ Add project to working sets   New...

Working sets: Select...

(?)   < Back   Next >   Finish   Cancel

Figure 4-4. *Creating a Java project*

27

3. Select File ➤ New ➤ Class to create a Java class
    (name: "Welcome"), as shown in Figure 4-5.
    Make sure you select "public static void
    main(String[] args)." Click on "Finish."

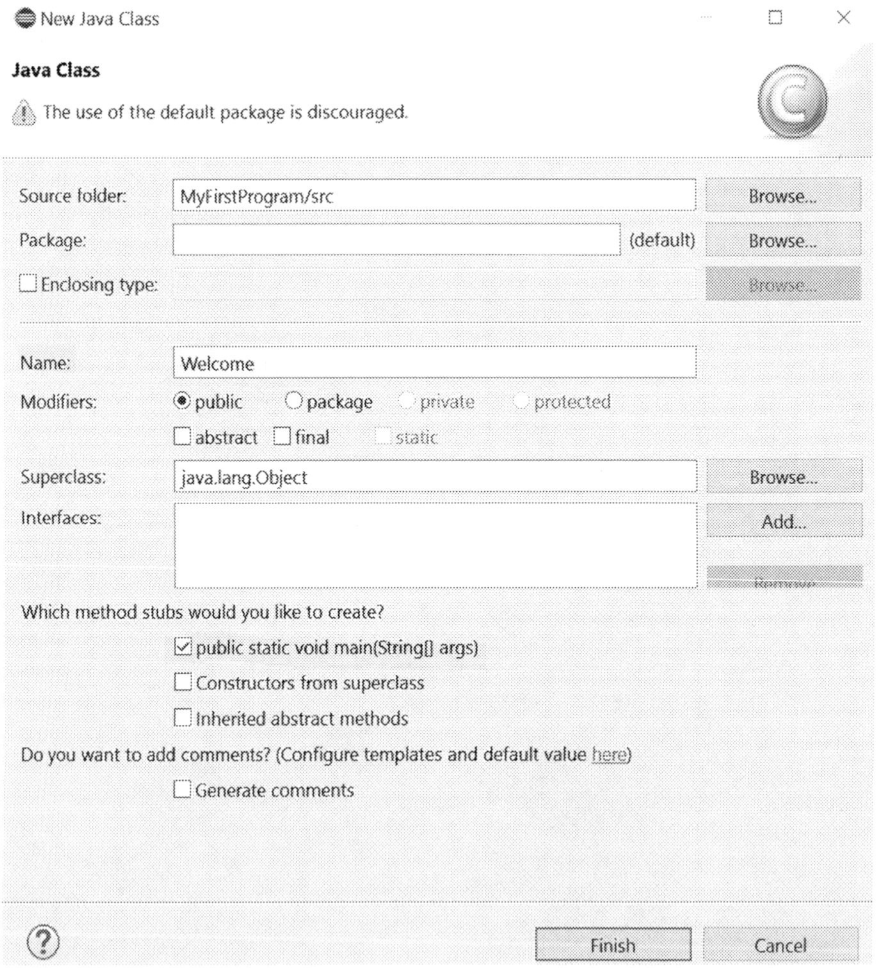

Figure 4-5. *Creating a Java class*

4. The Welcome class and the public static void main(String[] args) methods are automatically created, as shown in Figure 4-6. Then manually add the following output line:

```
System.out.println("Hello, friend,
you are welcome!");
```

5. Click on "Run" from the top menu bar, and then click on "Run" from its drop-down menu, we will see output text in the Console window as shown in Figure 4-6.

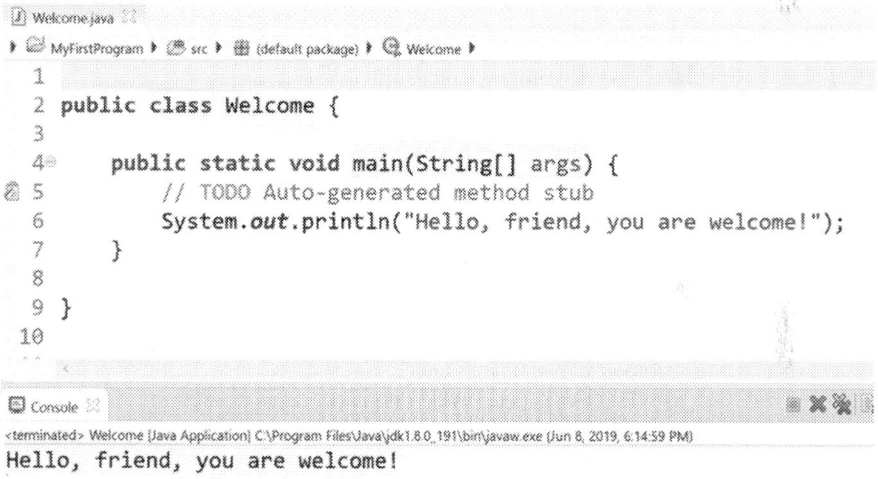

Figure 4-6. *Running the application*

Exploring Class and main()

As you saw in Chapter 3, a "class" is a template that describes the behavior that an object is supposed to show. You can create individual objects from the class. This is called "class instantiation." A class has local variables, instance variables, class variables, and a number of methods.

main() is a method name. When your Java program is executed, the runtime starts your program by calling its main() method first. The main() method is an entry point of your Java program.

Why Is It "public static void main(String[] args)"?

This is a convention designed by Java language and JVM (don't worry if some of this doesn't make sense, we'll come back to it later in the book).

- main is the name of the method;

- String[] args is the main() method input parameters as String array data type; the string values passed into the main() method are called arguments; they can be used as optional values to send to the program when it is started;

- void means there is no return data from the main() method call;

- public means the main() method is available for the JVM to call in order to start the execution of the whole program;

- static indicates that the main() method cannot be called with an object instance; in other words, the JVM can call it directly and does not have to create extra structures to call it.

If you change public to private, you will see the following error during runtime.

```
Error: Main method not found in class <your class name>, please
define the main method as:
    public static void main(String[] args)
or a JavaFX application class must extend javafx.application.
Application
```

Problems

1. What is the difference between the Something.java
    file and the Something.class file?

    (a) A .java file is a much larger binary file and a .class file is a
        smaller compressed version.

    (b) The .class file is for object-oriented programming and the
        .java file is for procedural programming.

    (c) A .java file contains code written in the Java language, and a
        .class file contains code written in the C++ language.

    (d) The programmer writes the .class file first, and the .java
        file is generated automatically later.

    (e) Something.java is a source code file typed by the
        programmer, and Something.class is a compiled executable
        class file that is run by the computer.

2. Which of the following method headers is correct?

    (a) public static foo void[]

    (b) public static void foo()

    (c) public void static foo{}

    (d) public void static foo()

    (e) public static foo()

CHAPTER 5

Variables

A variable is:

- the location of storage

- a container to store some kind of information for use at a later time

- retrievable by referring to a name that describes said information

There are different types of variables:

- Local variables

  They are variables only valid in local scope, such as inside a method or within a block of code.

- Class variables and instance variables

  We will see examples when we study basic class concepts in the later chapters. Class variables define data types for class fields and properties. When an object is created from a class, the class variables of the object become instance variables. Both class variables and instance variables are declared inside a class, but they don't belong to any method of the class.

© Ron Dai 2019

R. Dai, *Learn Java with Math*, https://doi.org/10.1007/978-1-4842-5209-3_5

- Method parameters

  Parameters are also variables used to pass values into a method from outside that method.

Defining a Variable Name

Variable names:

- cannot start with a number, or some special symbol, such as a quotation mark ("), or parentheses like ")", etc., but can start with an underscore (_) or dollar sign ($).

- cannot be any keyword that is already being used in the language, a.k.a. reserved word, for instance, if, else, and etc.

Example

Which of the following can be used in a Java program as identifiers? There is more than one answer.

1. ABC

2. B4

3. 24isThesolution

4. "hello"

5. AnnualSalary

6. _average

7. for

8. sum_of_data

9. first-name

10. println

Answer: 1, 2, 5, 6, 8, and 10

You may type simple code as shown in Figure 5-1 to check which strings are not qualified for variable names, because in Eclipse's Java code editor, all syntax errors are underlined in red.

```
public class Welcome {

    public static void main(String[] args) {

        int ABC = 1;
        int B4 = 1;
        int 24isThesolution = 1;
        int "hello" = 1;
        int AnnualSalary = 1;
        int _average = 1;
        int for = 1;
        int sum_of_data = 1;
        int first-name = 1;
        int println = 1;
    }
}
```

Figure 5-1. *Highlighting errors*

Different Types of Variables in Java

The types of variables will define the types of data as well as their size when stored in a variable. Java provides eight primitive types of data. Java also supports reference or object data types that are non-primitive types of data.

Primitive types:

- int

- long

- short

- `float`
- `double`
- `char`
- `byte`
- `boolean`

Reference types:

- `String`
- Object, Array, and so on

When we declare a variable in a program, we are actually reserving room in the computer's memory for operations. It is necessary to understand common data types and the memory space they occupy. This table shows a list of data types, their sizes in bits (and bytes), and the types of value they represent.

Type	Number of bits	Value
int	32 bits (= 4 bytes)	integer
short	16 bits (= 2 bytes)	integer
long	64 bits (= 8 bytes)	integer
byte	8 bits (= 1 byte)	integer
float	32 bits (= 4 bytes)	floating-point
double	64 bits (= 8 bytes)	floating-point
char	16 bits (= 2 bytes)	unicode character
boolean	See below	true / false

There are only two different values, true or false, for a Boolean data type. A single bit of room seems to be just a good fit. In fact, Java actually prepares for at least one byte's room for the Boolean data type, even though it only uses one bit of room. Put precisely, it is not clearly defined because it will be dependent on the virtual machine of the platform.

Assigning a Value to a Variable

Here is how to assign value or content to a specific type of variable:

```
int number1 = 3;
int number2 = 7;
int total = number1 + number2;
boolean flag = true;
String a = "welcome";
String b = "my friend";
String c = a + b;
```

You may then use the following method to display and validate the current values of the variables. For example, this line will display the current value of the string variable c:

```
System.out.println(c);
```

Lab Work

Referring to the first program we have created, utilize the System.out. println() statement as described; compile and run it; and then from the console window, verify the resulting value in each variable after each operation.

Basic math operation:

```
int number = 9 / 8;
double number = 9 / 8;
```

Math operations with order:

```
int number = 6 + 8 * 5;
int number = (6 - 8) * (5 + 3);
```

CHAPTER 6

First Algorithm

Today there are many different types of algorithms running on computers. An algorithm defines a set of instructions that a computer needs to follow to solve a specific problem. A smart and performant algorithm leads to an accurate and efficient working result.

Next is an example of creating an algorithm using real-world objects. Here we will look at how to exchange different kinds of water (fresh water and ocean water) between two containers of the same size. From common sense, we know to use a third empty container with the same size. Here is a series of actions you take to get this job done:

1. Pour fresh water from container A to container C (empty);

2. Pour ocean water from container B to container A;

3. Pour fresh water from container C to container B. Mission is accomplished.

© Ron Dai 2019
R. Dai, *Learn Java with Math*, https://doi.org/10.1007/978-1-4842-5209-3_6

You can see how water in each container was changed after each step from this table.

	Operation	Container A	Container B	Container C
Start	Introduce C	Fresh water	Ocean water	Empty
After step 1	A → C	Fresh water -> empty	Ocean water	Empty -> fresh water
After step 2	B → A	Empty -> Ocean water	Ocean water -> empty	Fresh water
After step 3	C → B	Ocean water	Empty -> fresh water	Fresh water -> empty

In many programs we will often run into situations when we need to set the value of a variable to that of another one. Let's apply the same logic we have learned from the last example to exchange values between two variables. In other words, we'll implement the algorithm we defined earlier.

Swapping Values Between Variables

Assume two integers, a = 5 and b = 4. We want to switch their values so that it will become a = 4 and b = 5. Following the order of operations listed in the next table, values in variables a and b will be switched over.

Step	Operation	a	b	c
0	int a = 5; int b = 4;	5	4	
1	int c = a;	5	4	5
2	a = b;	4	4	5
3	b = c;	4	5	5

Other Approaches

You may use other methods to swap values between the two integers without using a temporary variable. One method is by utilizing the + and − operators to exchange values:

$$a = a + b; \rightarrow \text{now } a = 9, \ b = 4$$

$$b = a - b; \rightarrow \text{now } a = 9, \ b = 5$$

$$a = a - b; \rightarrow \text{now } a = 4, \ b = 5$$

Successfully done!

CHAPTER 7

Input and Output

During runtime of a computer program, the program can ask the user to input data, read the user's input, and then show the user an output result. Scanner is the tool we use to implement the user interaction feature on the console window.

Importing java.util.Scanner

The Scanner utility class and its methods have been predefined in a package. We use the import statement to integrate the Scanner class with the program we are creating.

- Approach 1. Add line import java.util.Scanner; at the top of your Java code, and then add the following in your main() class:

  ```
  Scanner input = new Scanner(System.in);
  ```

- Approach 2. Type code Scanner input = new Scanner(System.in); in your Java code directly, and then use Eclipse's IntelliSense to choose the right fix. In other words, Eclipse will suggest that you add the import statement because it spots that you might need it.

  As a result, import java.util.Scanner; will be added at the top of the class.

© Ron Dai 2019
R. Dai, *Learn Java with Math*, https://doi.org/10.1007/978-1-4842-5209-3_7

Getting Input

There are several ways to read user input data from a program:

- nextLine(): read a string input

- next(): read a string input

- nextInt(): read an integer input

- nextFloat(): read a float number input

What is the difference between nextLine() and next()?

- next() reads the input only until the space.

  It cannot read two words separated by a space. And it places the cursor at the same line after reading the input stream, meaning it doesn't change the line.

- nextLine() reads the input until the end of the line ('\n').

  It will automatically move the scanner down after returning the current line.

Producing Output

System.out.println is a common way to display text in the console window. Developers often use it to read a user's input, provide general information to the user, and log information (to the console) during runtime in order to find out what is going on with key variables.

We often use the following special characters (i.e., escape characters) to control the output format:

- +: concatenates two strings;

- \n: a newline character;

\t: a tab key character that aligns text at the tab width;

\\: a backslash character;

\r: a carriage return character;

\" and \": double quote characters.

Here is an example:

```
System.out.println("This demonstrates " + "\"how to display a
table format\".\n");
System.out.println("123\t45\t6789\nab\tcde\tf");
```

This generates the following output:

```
This demonstrates "how to display a table format".

123     45      6789
ab      cde     f
```

Lab Work

Practice using the statement System.out.println() to:

1. display the string concatenation between two substrings "I am" and "a developer" using +

2. display a new line

3. display quotes using \" and \"

Example

Which of the following is the correct syntax to output a message?

1. System.out.println("Hello, world!");

2. System.println.out('Hello, world!');

45

3. `System.println("Hello, world!");`

4. `System.println(Hello, world!);`

5. `Out.system.println"(Hello, world!)";`

Answer: 1

Example

What is the output from the following statements?

```
System.out.println("\"Quotes\"");
System.out.println("Forward slashes \\//");
System.out.println("How '\"profound' \"\\\\" it is!");
```

Answer:

```
"Quotes"
Forward slashes \//
How '"profound' "\" it is!
```

Lab Work

What is the output from the following program? What if we replace the next() with nextLine()?

```
public class TestScanner {
        public static void main(String arg[]) {
                Scanner sc=new Scanner(System.in);
                System.out.println("enter string for c");
                String c=sc.next();
                System.out.println("c is "+c);
                System.out.println("enter string for d");
```

```
        String d=sc.next();
        System.out.println("d is "+d);
    }
}
```

Problems

1. What is the output produced from the following statements?

   ```
   System.out.println("name\tage\theight");
   System.out.println("Anthony\t17\t5'9\"");
   System.out.println("Belly\t17\t5'6\"");
   System.out.println("Bighead\t16\t6'");
   ```

2. What is the output produced from the following statements?

   ```
   System.out.println("\ta\tb\tc");
   System.out.println("\\\\");
   System.out.println("'");
   System.out.println("\"\"\"");
   ```

3. Write a program in Java to print the following:

   ```
   \/
   \\//
   \\\///
   ```

CHAPTER 8

Loop Structure – for Loop

Simply put, the loop structure repeatedly does something until the state is changed (see Figure 8-1).

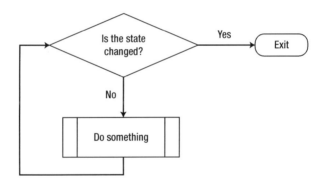

Figure 8-1. *The for loop structure*

Example

Here is an example:

```
for (int i = 0; i < 100; i++) {
        <do something>
}
```

There are three key elements in the `for` statement:

- `int i = 0`: declare a counter variable and assign an initial value to it;

- `i < 100`: define the condition to continue with the `for` loop; as long as this condition is true the loop will run, when it is not true, we stop and exit from the `for` loop;

- `i++`: increment the counter value; `i++` is the same thing as `i = i + 1.`

So, how many times will the "<do something>" be repeated in the code above?

Lab Work

1. Use the `for` loop to output "Hello!" 10 times.

2. Use the `for` loop to print out all integers from 1 to 25, inclusive.

3. Print out all integers from 1 to 25.

4. Output all even numbers from 3 to 99.

The for Loop Formula

The math model behind the `for` loop is actually an arithmetic sequence:

```
for (int counter=firstTerm;
     counter <= lastTerm;
     counter=counter + difference) {
     ......
}
```

The nth term in the `counter` series is equal to:

firstTerm + difference × (n – 1)

Finding the "for Loop" Formula for an Arithmetic Sequence

As an example, here is a list of numbers:

-4, 5, 14, 23, 32, 41, 50, 59, 68, 77, 86.

It follows an arithmetic sequence.

firstItem = -4

lastItem = 86

difference = 5 – (-4) = 9

Translate this to a `for` loop:

```
for(int i=-4; i <= 86; i=i+9) { ...... }
```

It will iterate through every single number in the list.

Math: Counting Strategically

You may finger count, but that will not work when you have an extremely large amount of numbers in the series. The right approach is to prepare these numbers by reorganizing them. The purpose is to find a good pattern so that we can count systematically.

Finding a pattern here is to figure out a basic formula representing every number in the series.

Look at this example: 3, 4, 5, 6, ..., 100, so we know the total count of numbers is:

$$100 - 3 + 1 = 98.$$

A common method is to convert the number series to something more straightforward. If we subtract 2 from every number in the series, we get 1, 2, 3, 4, ..., 98. We now know the count is 98. And, the formula representing every number will be $x(i) = i + 2$, $(i=1, 2, ..., 98)$.

What about 5, 8, 11, 14, ..., 101? How do you use the "for loop" to print it out?

It looks more complicated than the previous one, but you can try the same approach.

1. Subtract 5 from every number; it becomes 0, 3, 6, 9, ..., 96

2. Divided by 3, it then becomes 0, 1, 2, 3, ..., 32

3. It is not hard to count from 0, one by one up to 32. The total count is 33.

4. The general term for the i-th number in the series will be $x(i) = 3 * i + 5$, $(i=0, 1, 2, ..., 32)$.

Now, go back to the for loop construction, and it is obvious the answer should be something like this:

```
for (i=0; i <= 32; i++) {
        System.out.println(3 * i + 5);
}
```

Lab Work

- Write a for loop to produce the following list of numbers:

  1 4 9 16 25 36 49 64 81 100

  (Hint: watch for a common pattern.)

Example

What is the output of the following sequence of loops?

```
for (int i = 1; i <= 2; i++) {
    for (int j = 1; j <= 3; j++) {
        for (int k = 1; k <= 4; k++) {
            System.out.print("*");
        }
        System.out.print("!");
    }
    System.out.println();
}
```

It prints out the following:

****!****!****!
****!****!****!

The external for loop (marked as "1") has two iterations, so the output will have two lines, by println().

The middle for loop (marked as "2") has three iterations, so it will print out 2 x 3 = 6 "!" in total by print().

The internal for loop (marked as "3") has four iterations, so it will print out 2 x 3 x 4 = 24 "*" in total by print().

Lab Work

- Write a method exp() to compute an exponential result, given the input of a base number and a power (a.k.a. an exponent number). For example, exp(3, 4) returns 81. The restriction is that the base and exponent numbers are non-negative.

Problems

1. What is the output of the following sequence of loops?

```java
for (int i = 1; i <= 2; i++) {
    for (int j = 1; j <= 3; j++) {
        System.out.print(i + ""*"" +
            j + ""= "" + i * j + ""; "");
    }
    System.out.println();
}
```

2. Write a for loop to produce the following list of numbers:

   5 10 17 26 37 50

3. Write a for loop to produce the following list of numbers:

   1 8 27 64 125

4. Write a for loop to produce the following list of numbers:

   -1 0 7 26 63 124

5. Use for loops to produce the following output:

    *****

    *****

    *****

    *****

6. Write for loop code to output the following:

    1

    22

    333

    4444

CHAPTER 9

Loop Structure – while Loop

This is another way for the loop structure (Figure 9-1).

```
while(i == 0) {
    <do something>;    //the variable "i" may be
    updated in this code block.
}
```

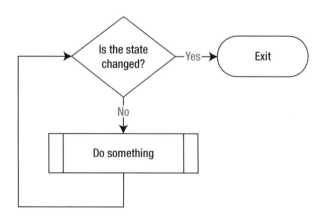

Figure 9-1. The while loop

© Ron Dai 2019
R. Dai, *Learn Java with Math*, https://doi.org/10.1007/978-1-4842-5209-3_9

The state may be updated within the "Do something" process.

Question: What will happen if the state is never changed?

Answer: it will be an "infinite loop," meaning the program will run forever until it crashes or is terminated by the user.

Note   You may have noticed the `//` next to the `<do something>` line. This identifies a comment left by the developer. You should use comments to annotate your code so it's easier to understand for future readers (who could be you, so be kind to your future self). The compiler will ignore all comments when compiling your Java programs.

Example

How many times will the loop execute its body?

```java
int x = 1;
while (x < 100) {
    System.out.print(x + " ");
    x += 10;
}
```

Answer: ten times (when x = 1, 11, 21, 31, 41, 51, 61, 71, 81, 91)

Example

How many times will the loop execute its body?

```java
int max = 10;
while (max < 10) {
    System.out.println("count down: " + max);
    max--;
}
```

Answer: zero.

Both the `for` loop and `while` loop are loop structures to accomplish a repetitive job (i.e., <do something>). The `for` loop has provided an easy way to assign an initial counter value and to define how the counter value is incremented (or decremented) within the same line of code, while the `while` loop requires the user to define them in separate lines. The following two examples have an equivalent functionality.

```
for(int i = 0; i < 10; i++) {
        <do something>
}

int i = 0;
while(i < 10) {
        <do something>
        i++;
}
```

There is a good reason why we need the `while` loop option, in addition to the `for` loop. This example shows one of many circumstances when we prefer the `while` loop over the `for` loop.

```
        boolean flag = true;
        while(flag) {
                < commit planned operations, during which time
                the flag may be updated upon a certain codition
                change, e.g. the operation is completed, or
                failed for some reason.>
        }
```

The do-while Loop

Java also provides a do-while loop structure:

```
do {
    <do something>
} while (expression);
```

The difference between do-while and while is that do-while evaluates its Boolean expression at the bottom of the loop instead of the top. Therefore, the statements within the do block (a.k.a. <do something>) are always executed at least once. You may try the following program to see a demo.

```
class DoWhileDemo {
    public static void main(String[] args){
        int count = 1;
        do {
            System.out.println("Count is: " + count);
            count++;
        } while (count < 1);
    }
}
```

Lab Work

1. Use the while loop to output "Hello!" 10 times.

2. Use the while loop to print out all integers from 1 to 25, inclusively.

3. Explain what the following code snippet is trying to do.

```
int n = 5;
while (n == 5) {
        n = n + 1;
        System.out.println(n);
        n--;
}
```

Problems

1. How many times will the loop execute its body?

```
int x = 250;
while (x % 3 != 0) {
    System.out.println(x);
}
```

2. How many times will the loop execute its body?

```
int x = 2;
while (x < 200) {
    System.out.print(x + " ");
    x *= x;
}
```

3. How many times will the loop execute its body?

```
String word = "a";
while (word.length() < 10) {
    word = "b" + word + "b";
}
System.out.println(word);
```

4. How many times will the loop execute its body?

```
int x = 100;
while (x > 1) {
    System.out.println(x / 10);
    x = x / 2;
}
```

5. Given the static method runWhileLoop(), what is its output when x = 10? You may want to copy this method to your test class to try.

```
public static void runWhileLoop(int x) {
    int y = 1;
    int z = 0;
    while (2 * y <= x) {
        y = y * 2;
        z++;
    }
    System.out.println(y + " " + z);
}
```

CHAPTER 10

Logical Control Structures

Very similarly to how we describe it verbally when there is a logical conversation, the if and the if/else in programming languages are common structures to make conditional decisions and choose corresponding execution paths (Figure 10-1).

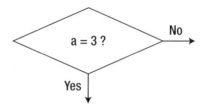

Figure 10-1. *The if structure*

```
if (a == 3) {
        <do something>
}
```

This is similar to, but not completely the same, as:

```
if (a == 3) {
        <do something>
} else {
        <do something else>
}
```

© Ron Dai 2019
R. Dai, *Learn Java with Math*, https://doi.org/10.1007/978-1-4842-5209-3_10

Figure 10-2 is the whole workflow of the if/else logical control structure.

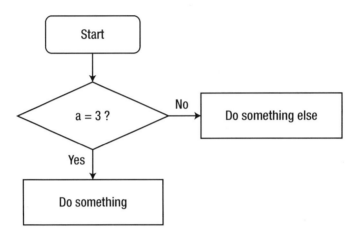

Figure 10-2. *The if/else structure*

Conditional Operators

The conditional operators listed in the next table are frequently used in conditional statements. For example, we use a == 3 to evaluate "is a equal to 3?" in a conditional statement. Java uses six different conditional operators to express a relationship between two operands. The results of the expressions are either true or false, a Boolean value, which determines the "yes" or "no" path of execution.

Conditional Operators	Description
==	Is equal to
>	Is greater than
>=	Is greater than, or equal to
<	Is less than
<=	Is less than, or equal to
!=	Is not equal to

Example

Which of the following if statement headers uses the correct syntax?

(a) if x = 10 then {

(b) if (x equals 42) {

(c) if (x => y) {

(d) if [x == 10] {

(e) if (x == y) {

Answer

e

Example

Given the following method, what is the output from whatIsIt(9, 4)?

```java
public static void whatIsIt(int x, int y) {
    int z = 4;
    if (z <= x) {
        z = x + 1;
    } else {
        z = z + 9;
    }
    if (z <= y) {
        y++;
    }
    System.out.println(z + " " + y);
}
```

Answer

10 4

Lab Work

1. Define an integer variable and assign value "3" to it.

2. Use an if statement to output "Hello" when the integer variable is assigned number 3.

3. Use an if/else statement to output "Goodbye" when any number other than 3 is assigned to the integer variable.

4. Is there anything wrong with the following code?

```
int n = 4;
if (n >= 3) {
        System.out.println("Hello!");
}
if (n == 4) {
        System.out.println("Hello again!");
}
```

5. Use an if/else statement to implement the following requirements:

   • Output "less than 3" when the number is smaller than 3

   • Output "equals 3" when the number is 3

   • Output "greater than 3" when the number is bigger than 3

6. Input an integer number, and then,

   • Output "The number is greater than 6" when the input number is bigger than 6

   • Output "The number is smaller than 6" when the input number is smaller than 6

7. Explain what the following code snippet is trying
    to do:

```
Scanner scan = new Scanner(System.in);
int n = scan.nextInt();
if (n > 6) {
        if (n > 8) {
                System.out.println("n is greater
                than 8");
        }
        else {
                System.out.println("n is greater
                than 6, but n is smaller than 9");
        }
}
```

Sometimes you may need to use a logical combination of multiple
"true or false" conditions. Let's introduce another concept here, in terms of
"Logical Operators."

Logical Operators

Math: Logical Operators
Logical operators and logical operations:

&& → AND relation

|| → OR relation

! → NOT relation

A && B → indicates only when both A and B are
true, the result is true. For example, in (x > 3 && x < 5),
A is "x > 3", B is "x < 5".

A || B → indicates when either A or B is true, the result is true.

!A → indicates when A is true, the result is false; when A is false, the result is true. In the example of "!(x > 0)", A is "x > 0".

In all of these examples, A and B are expressions or Boolean variables.

(A && B)	A=true	A=false
B=true	True	False
B=false	False	False

Summary - Result is true, only when both A and B are true. Otherwise, result is false.

(A || B)	A=true	A=false
B=true	True	True
B=false	True	False

Summary - Result is false, only when both A and B are false. Otherwise, result is true.

A final example of some properties of these operators:

- $(x < 0 \, || \, x > 0) \longleftrightarrow (x \, != 0)$

- $!(x == 0 \, || \, y == 0)$ is equivalent to $(x \, != 0 \, \&\& \, y \, != 0)$

- $!(x > 3 \, \&\& \, x < 5)$ is equivalent to $(x >= 5 \, || \, x <= 3)$

Using a Venn diagram will help us analyze some type of logical problems.

Math: Analyzing Logical Problems

A Venn diagram is a visualization method to reveal logical relations among data sets.

In Figure 10-3, the overlap area between circle A and circle C is in area B.

if we define A = { x , y | x = 0 }, C = { x, y | y = 0 }, then B = { x=0; y=0 }.

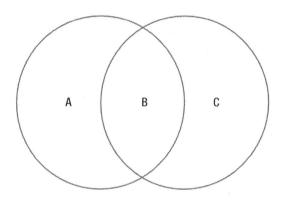

Figure 10-3. *A Venn diagram*

Lab Work

1. Figure out the output of the following program.

```java
public class LogicalOperation {
        public static void main(String args[]) {
                boolean a = true;
                boolean b = false;
                System.out.println("a && b = "
                + (a&&b));
                System.out.println("a || b = "
                + (a||b) );
```

```
                        System.out.println("!(a && b) = "
                        + !(a && b));
            }
}
```

2. Write a static method called quadrant that takes as
    parameters a pair of integer numbers representing
    an (x, y) point on the Cartesian coordinate system.
    It returns the quadrant number (i.e., 0,1,2,3,4, see
    picture) for that point.

Below are sample calls on the method.

Call	Value Returned
quadrant(12, 17)	1
quadrant(-2, 3)	2
quadrant(-15, -3)	3
quadrant(4, -42)	4
quadrant(0, 3)	0

Problems

1. Translate the following English statements into
    logical expressions:

    (a) z is odd.

    (b) x is even

    (c) y is positive.

    (d) Either x or y is even.

(e) y is a multiple of z.

(f) z is not zero.

(g) y is a positive number, and y is greater in magnitude than z.

(h) x and z are of opposite signs.

(i) y is a nonnegative one-digit number.

(j) z is nonnegative.

2. Given the following variable declarations: `int x = 4;`
    `int y = -3; int z = 4;`

    What are the results (True or False) of the following expressions?

    `x == 4 x == y`

    `x == z y == z`

    `x + y > 0 x - z != 0`

    `y * y <= z y / y == 1`

    `x * (y + 2) > y - (y + z) * 2`

CHAPTER 11

Errors and Tips

The following is a list of common coding mistakes that beginners can easily make. It will help you overcome initial coding barriers if you are mindful of these error patterns.

- Missing half of curly braces {}.

  (It should always come with a pair.)

- Missing half of parentheses ( ).

  (It should always come with a pair too.)

- Missing semicolon ; at the end of each line.

- Use one = signs in a condition check.

  (Correct way is ==.)

- Assigning a value to a variable that is not defined.

  (Correct way is assigning a value to the variable only after it has been defined.)

- Defining the same variable more than once.

  ```
  int i;
  ......
  int i = 3;
  ......
  ```

© Ron Dai 2019
R. Dai, *Learn Java with Math*, https://doi.org/10.1007/978-1-4842-5209-3_11

- Forgetting to increment/decrement the counter inside a loop structure.

- Incorrect signature of main function.

  It should be `public static void main(String[] args`, so pay attention to `public static void main`, and `String[] args`).

- No output on console window – missing output line, for example, `System.out.println()`.

- Differences between the variable name and the string:

  - Variable name a is a string.

    ```
    string a;
    ```

  - Variable name a is a string with value "a".

    ```
    string a = "a";
    ```

- Mistakenly resetting value in an aggregator:

  ```
  for (int i=1; i<n; i++) {
          int sum = 0;
          sum += i;
  }
  ```

This program is to sum up all numbers from 1 up to n. To correct the mistake, the line `int sum = 0` needs to be moved out of the loop structure, right before the `for` line.

Programming Tips

- How to increase/decrease the font size on text editors:

  Use Ctrl + or Ctrl -.

  On macOS that would be ⌘+ and ⌘-

  Setting: Preferences => General => Keys

- How to comment out a code section in Eclipse:

  Select code block

  Press CTRL and "/" (simultaneously)

  On macOS, it is Command + "/"

- How to make comments in a code block:

  You may use the Java Comments feature to briefly explain what a specific line of code or a code block in your program is doing so that other people can understand your implementation ideas. There are basically two ways you can write Java Comments among lines of code.

  1. Use "//" as a prefix to write a statement in one line, for example:

     // count is a variable to track the total number of clicks

     int count = 0;

2. Use "/*" and "*/" to write multiple lines of comments, for example:

    /*

    This block of code tries to find a maximum price value (in dollars) from the specified group of products:

    */

- Watch out "for the colored underline" when coding:

    Red line – error message: syntax, etc.

    Orange line – warning message

Handling Exceptions

So far in this chapter we have explained how to avoid making mistakes that will be caught by the compilation error detection process. What about those errors during runtime? In Java, we use the following structure to capture them and handle these conditions separately. This is called exception handling.

```
try {
        <main instruction code to execute>
        ......
} catch(IllegalArgumentException ex) {
        <exception handling steps>
}
```

The code that could cause an error at runtime goes in the try block, and the code to respond to an error at runtime goes in the catch block. Here the catch block responds only to errors that throw an

IllegalArgumentException. You can specify multiple catch blocks to respond to different types of exceptions thrown at runtime because you may want to respond differently to different types of runtime errors. Finally, you can throw an error on purpose if your code detects some problem at runtime. You'll see how to do this in Chapter 17.

Problems

1. The following program contains three errors. Correct the errors and submit a working version of the program.

    ```
    public MyProgram {
        public static void main(String[]
        args) {
            System.out.println("This is a test
            of the")
            System.out.Println("alarming system.");
            System.out.printLn("Thank you for your
            attention!")
        }
    }
    ```

2. The following program contains four errors. Correct the errors and submit a working version of the program.

    ```
    public class FriendMessage {
        public static main(string[] args) {
            System.out.println("Speaking plum");
            System.out.println("and eat);
        }
    ```

3. The following program contains at least 10 errors. Correct the errors and submit a working version of the program.

```
public class Many Errors {
    public static main(String args) {
        System.println(Hello, buddy!);
        message()
    }

    public static void message {
        System.out println("This program
        cannot ";
        System.out.println("have any
        "errors" in it");
    }
}
```

CHAPTER 12

Java Basics Summary

In this chapter we will summarize what we have learned so far in the Java basics area and point out some common mistakes.

General Rules

How to Define a Variable Name

Variable name string

- case-sensitive

- okay to include numbers 0 to 9

- okay to have underscore _ or dollar sign $

Variable name string:

- cannot start with number

- cannot use reserved words, such as `for`, `class`, `void`, `if`, `else`, etc.

How to Output in Console

- `System.out.print(<string + value>);`

- `System.out.println(<string + value>);`

© Ron Dai 2019
R. Dai, *Learn Java with Math*, https://doi.org/10.1007/978-1-4842-5209-3_12

How to Listen to Input in Console

```
Scanner input = new Scanner(System.in);
input.nextLine();
input.next();
input.nextInt();
input.nextFloat();
```

How to Repeat an Operation

```
for(<initial state>; <condition check>; <increment/decrement
count>) {
  <do something>;
}

while(<condition check>) {
  <do something>
}
```

or,

```
do {
      <do something>
}
while(<condition check>)
```

How to Control a Conditional Operation

```
if (<condition check>) {
  <do something>;
}
```

```
if (<condition check>) {
  <do something>;
} else {
  <do something else>;
}

if (<condition check>) {
  <do something>;
}
else {
  <do something else with the nested if/else statement(s)>;
}
```

Basic Coding Structure

```
public class MyClass {
      public static void main(...) {
            myMethod();
      }
      private static void myMethod(...) {
            for (... ; ... ; ...) {
                  ......;
            }
            if (...) {
                  ......;
            }
            else {
                  ......;
            }
      }
}
```

81

Curly Braces

- Always come with a pair of curly braces: "{ . . . . . . }"

- Always come first with: "{", and then "}"

- Common patterns (two pairs of open/close as an example)

  - {{}} → open, open, close, close

  - {}{} → open, close, open, close

  - {}}{ ← wrong!

  - }{}{ ← wrong!

- What is the basic rule of these patterns?

  - At beginning, start with "{"

  - Finally, end with "}"

  - Never has more "}" than "{"

Lab Work

1. What is the output produced from the following program?

```
public class StoryOfMethods {
        public static void main(String[] args) {
                method1();
                method2();
                System.out.println("Done with main.");
        }
```

```java
public static void method1() {
        System.out.println("This is from
        method1.");
}
public static void method2() {
        System.out.println("This is from
        method2.");
        method1();
        System.out.println("Done with method2.");
    }
}
```

2. What is the output produced from the following program?

```java
public class OrderOfFunctions {
        public static void main(String[] args) {
                second();
                first();
                second();
                third();
        }
        public static void first() {
                System.out.println("Inside
                the first function.");
        }
        public static void second() {
                System.out.println("Inside the
                second function.");
                first();
        }
```

```java
public static void third() {
        System.out.println("Inside the third
        function.");
        first();
        second();
    }
}
```

CHAPTER 13

Java Basics Projects

A.   Write code to print out the following graph.

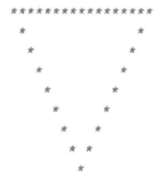

B.   Write code to draw the following shape.

© Ron Dai 2019

R. Dai, *Learn Java with Math*, https://doi.org/10.1007/978-1-4842-5209-3_13

C. Write code to make this triangle.

```
*
**
***
****
*****
******
*******
```

D. Write a function to check if an integer is divisible by another integer.

For example:

(1) Given input 10, 5, output is "yes, 2";

(2) Given input 11, 2, output is "no";

E. Write Java code to draw the Christmas tree.

```
      **
      **
     ****
    ******
   ********
  **********
 ************
**************
***************
****************
     ****
      **
      **
```

F. Write code to find all factors for a given positive integer.

For example, when a user inputs "10," your program should output 1, 2, 5, 10

G. Write a method that accepts a month (i.e., an integer between 1 and 12) as a parameter and returns the number of days in that month in the current year.

H. Write code to produce a product of two numbers that are inputs from the console:

Please enter two numbers for multiplication

Next number --> 7

Next number --> 15

Product = 105

I. Write a method to examine whether a positive integer is a prime number or not.

J. Write a method to convert an integer number to a string of that number's representation in binary. For example, given a parameter "19," it should return "10011".

CHAPTER 14

Java Basics Solutions

Here are the solutions from the previous chapters in the Java Basics part of the book.

Chapter 5: Variables

(1)   Signal (on/off) of the digital electronics

(2)   OOP; Class-based; WORA

(3)   (e)

(4)   (b)

Chapter 7: Input and Output

(1)

```
name      age     height
Anthony   17      5'9"
Belly     17      5'6"
Bighead   16      6'
```

(2)

```
a    b    c
\\
'

" " "
```

© Ron Dai 2019
R. Dai, *Learn Java with Math*, https://doi.org/10.1007/978-1-4842-5209-3_14

Chapter 8: Loop Structure –For Loop

(1) Inner loop has three iterations without carriage return; Outer loop has two iterations.

(2) After subtracting one, you will find a clear pattern.

(3) Cubic number pattern

(4) Adding one to all numbers

(5) Nested loops

Chapter 9: Loop Structure – While Loop

(1) Forever (dead loop)

(2) Three times, x = 2, 4, 16

(3) Five times, final output is "bbbbbabbbbb"

(4) Six times, when x = 100, 50, 25, 12, 6, 3

(5) 8 3 ← There is a space in the middle

Chapter 10: Logical Control Structures

(1)

    (a) $z \% 2 == 1$

    (b) $x \% 2 == 0$

    (c) $y > 0$

    (d) $(x \% 2 == 0) \,||\, (y \% 2 == 0)$

    (e) $y \% z == 0$

    (f) $z \,!= 0$

    (g) $(y > 0 \,\&\&\, y > z \,\&\&\, z >= 0) \,||\, (y > 0 \,\&\&\, y > \text{-}z \,\&\&\, z < 0)$

    (h) $(x > 0 \,\&\&\, z < 0) \,||\, (x < 0 \,\&\&\, z > 0)$

    (i) $(y >= 0 \,\&\&\, y < 10)$

    (j) $z >= 0$

PART II

Java Intermediate

Readers should have completed Part I: Java Basic prior to this part. Part II focuses on how we learn Java programming and integrates basic mathematical concepts.

Also in Part II, we demonstrate how to apply Java programming to math problem solving through many practical examples.

Readers will have the opportunity to witness how Java programming becomes a powerful tool in our experimental work.

I am sure you will be excited to find many intriguing examples of applications throughout this part. Although this book doesn't touch every single detail, it will cover the basic concepts of class and object-oriented programming so that beginners are able to build a good foundation.

CHAPTER 15

Wright Brothers' Coin Flip Game

Programming helps us to understand and explain many complicated problems. You can find an interesting online video, "The coin flip conundrum," which tells a historic story and explains a probability problem solution using an analytic approach. The story is about the Wright brothers, Orville and Wilbur, who played a coin flip game to determine who should start the new flight experimentation first. They flipped a coin continuously, until Orville got double heads consecutively, or Wilbur received a head and a tail in a "neighboring" sequence. While the video used probability plus algebraic concepts to calculate the winning edge for the Wright brothers, we are going to try the experimentation with Java programming. The following method simulates the Wright brothers' game and analyzes their results (note the helpful comments, marked by //, so we can read the code more easily).

```
private static int count_a, count_b, count_ab = 0;
private static int totalsteps_a, totalsteps_b = 0;

/// whoever gets below pattern first wins, or tie if both of
them reach targeted patterns at the same round
/// a: HH wins; b: HT wins; Use boolean 'true': head, 'false':
tail
```

© Ron Dai 2019
R. Dai, *Learn Java with Math*, https://doi.org/10.1007/978-1-4842-5209-3_15

```java
public static void flipCoin()
{
        Random r = new Random();
        /// initial value, or first round result
        boolean current_a = r.nextBoolean();
        boolean current_b = r.nextBoolean();

        boolean win_a = false;
        boolean win_b = false;
        int round = 1;
        while(true) {
                round++;
                boolean next_a = r.nextBoolean();
                boolean next_b = r.nextBoolean();
                if (current_a && next_a) {
                        win_a = true;
                }
                if (current_b && !next_b) {
                        win_b = true;
                }
                if (win_a && win_b) {
                        System.out.println("Both WIN!
                        - round: " + round);
                        count_ab++;
                        totalsteps_a += round;
                        totalsteps_b += round;
                        break;
                }
                if (win_a && !win_b) {
                        System.out.println("A WIN! - round: "
                        + round);
                        count_a++;
```

```
                totalsteps_a += round;
                break;
        }
        if (!win_a && win_b) {
                System.out.println("B WIN!
                - round: " + round);
                count_b++;
                totalsteps_b += round;
                break;
        }
        current_a = next_a;
        current_b = next_b;
    }
}
```

Using the following main method, we can collect samples and get the statistical summary.

```
public static void main(String[] args) {
        final int MAX = 10000;
        for(int i=0; i < MAX; i++) {
                flipCoin();
        }
        System.out.println("Summary");
        System.out.println("Total samples: " + MAX);
        System.out.println("Winning counts: a - " + count_a + ";
        b - " + count_b + "; ab - " + count_ab);
        int probability_a = count_a * 100 / (count_a + count_b);
        int probability_b = count_b * 100 / (count_a + count_b);
        System.out.println("Winning probability: HH=" +
        probability_a + "%; HT=" + probability_b + "%.");
        double average_a = totalsteps_a / (count_a + count_ab);
```

```
        double average_b = totalsteps_b / (count_b + count_ab);
        System.out.println("Average rounds to win: HH=" +
        average_a + "; HT=" + average_b + ".");
}
```

After we run many experiments with different parameters, we learn what is actually going on and can then reach the conclusion that Wilbur would have a significantly higher chance (roughly 62% vs. 37%) to win the bet. The output should look similar to what follows.

```
. . . . . . . . .
B WIN! - round: 3
A WIN! - round: 2
A WIN! - round: 3
B WIN! - round: 2
B WIN! - round: 4
A WIN! - round: 4
B WIN! - round: 2
B WIN! - round: 2
B WIN! - round: 3
B WIN! - round: 2
A WIN! - round: 2
A WIN! - round: 2
Both WIN! - round: 2
B WIN! - round: 3
Summary
Total samples: 10000
Winning counts: a - 3213; b - 5386; ab - 1401
Winning probability: HH=37%; HT=62%.
Average rounds to win: HH=2.0; HT=3.0.
```

CHAPTER 16

Pythagorean Triples

The Pythagorean Theorem is well known among elementary to middle school students, given its elegantly looking equation that applies to all right triangles.

Math: Pythagorean Triples

Inside a right triangle, a and b are the two legs, and c is the hypotenuse.

$$a^2 + b^2 = c^2$$

When a, b, and c are positive integers satisfying the Pythagorean Theorem, (a, b, c) are called "Pythagorean Triples". Obviously (3, 4, 5) is the first Pythagorean triple, followed by (5, 12, 13), (6, 8, 10), and so forth. The number of Pythagorean triples is infinite.

Example

So, how do we find all the Pythagorean triples below 100?

Answer

We can multiply (3, 4, 5) with any integer number to get (6, 8, 10), (9, 12, 15), …, (57, 76, 95). We can use (5, 12, 13) as another base triple to get (10, 24, 26), …, (35, 84, 91). And so on.

But this approach will require us to first find out all the base Pythagorean triples. Thus, we would essentially have to check every positive integer below 100 for a, and then figure out b and c, assuming a < b < c. By the way, a = b will not be possible. However, with a programming approach, it is no longer a challenging math problem.

© Ron Dai 2019
R. Dai, *Learn Java with Math*, https://doi.org/10.1007/978-1-4842-5209-3_16

This is the method to find out all the possible triples (a, b, c) satisfying the Pythagorean Theorem.

```java
private static int allPythagoreanNumbers(int upperBound) {
    int count = 0;
    for(int a = 1; a < upperBound; a++) {
        for(int b = a; b < upperBound; b++) {
            for(int c = b; c < upperBound; c++) {
                if (a * a + b * b == c * c) {
                    System.out.println
                    ("("+a+", "+b+",
                    "+c+")");
                    count++;
                }
            }
        }
    }
    return count;
}

public static void main(String[] args) {
    System.out.println("Total count: " +
    allPythagoreanNumbers(100));
}
```

It will output something like what follows:

(3, 4, 5)

(5, 12, 13)

(6, 8, 10)

(7, 24, 25)

(8, 15, 17)

$(9, 12, 15)$

$(9, 40, 41)$

$(10, 24, 26)$

$(11, 60, 61)$

$(12, 16, 20)$

$(12, 35, 37)$

$(13, 84, 85)$

$(14, 48, 50)$

$(15, 20, 25)$

$(15, 36, 39)$

$(16, 30, 34)$

$(16, 63, 65)$

$(18, 24, 30)$

$(18, 80, 82)$

$(20, 21, 29)$

$(20, 48, 52)$

$(21, 28, 35)$

$(21, 72, 75)$

$(24, 32, 40)$

$(24, 45, 51)$

$(24, 70, 74)$

$(25, 60, 65)$

$(27, 36, 45)$

$(28, 45, 53)$

$(30, 40, 50)$

$(30, 72, 78)$

$(32, 60, 68)$

$(33, 44, 55)$

$(33, 56, 65)$

$(35, 84, 91)$

$(36, 48, 60)$

$(36, 77, 85)$

$(39, 52, 65)$

$(39, 80, 89)$

$(40, 42, 58)$

$(40, 75, 85)$

$(42, 56, 70)$

$(45, 60, 75)$

$(48, 55, 73)$

$(48, 64, 80)$

$(51, 68, 85)$

$(54, 72, 90)$

$(57, 76, 95)$

$(60, 63, 87)$

$(65, 72, 97)$

Total count: 50

This is just one of many demonstrations of how we can use programs to solve problems.

Problems

1. In the example, we used three for-loops to iterate a, b, c from 1 through 99. How do you improve it by reducing to two for-loops?

2. Using the idea from the example, how do we find out all the Pythagorean primes smaller than 100? Pythagorean primes are explained below.

Math: Pythagorean Primes

Pythagorean primes are the sum of two squares. And, it needs to be in form of 4n + 1, where n is a positive integer. Examples of Pythagorean primes are 5, 13, 17, 29, 37 and 41.

Hine   Take advantage of the example code and see how to make small changes to find a solution.

CHAPTER 17

Strong Typed Programming

As we have learned at the beginning of this book in Part I, the Java programming language has defined the integer, double, Boolean, String, etc., as basic types. In Java, we cannot assign any value to a variable without defining the variable with a type beforehand. Only after a variable has been defined clearly by a type—for example, integer—are we then allowed to assign an integer value to it and start using it as an integer in the calculation. Once a variable's type is defined, it cannot be assigned a value with a different type, logically speaking. For instance, if a variable is defined as Boolean, it cannot be assigned an integer value. Otherwise, we will get a type mismatch compilation error.

Type Casting

However, if a variable is defined as a double, how do we assign an integer value to it? Let's do some experimentation.

```java
public static void main(String[] args) {
    double a = 5;
    System.out.println(a);
    double b = 3 * 5;
    System.out.println(b);
```

```
double x = 5 / 3;
System.out.println(x);
double y = (double)(5 / 3);
System.out.println(y);
double z = (double)5 / 3;
System.out.println(z);
double t = 5.0 / 3;
System.out.println(t);
double u = 5 / 3d;
System.out.println(u);
    }
```

This is the output:

```
5.0
15.0
1.0
1.0
1.6666666666666667
1.6666666666666667
1.6666666666666667
```

The following patterns are those we have learned from this experiment:

- When the integer 5 is assigned to a double typed variable, the variable will get an equivalent value with a decimal point presentation, that is, double value 5.0.

- When an integer is divided by another integer, the result follows the same integer type, for example, 5 / 3 = 1. But when the fraction "5 / 3" is assigned to a double typed variable, the resulting value will be automatically converted to double value 1.0.

- (double)(5 / 3) converts an integer result of (5/ 3) to a double value. This is called type casting. The result is 1.0, a double value.

How do we produce a precise value from 5 / 3? The trick is to use (double)5 / 3, instead of 5 / 3. The outcome of (double)5 / 3 is a double value. Or, you may use 5.0 / 3 to generate the same outcome. Another way is 5d / 3, or 5 / 3d. The outcomes of both expressions are the same double value.

Math: Slope of a Line

In the x-y 2D Cartesian coordinates system, the slope of a line between points (x1, y1) and (x2, y2) is equal to (y2 - y1) / (x2 - x1).

Example

Implement a public method called double getSlope(), which returns the slope of a line. If the two points have the same x-coordinates, the denominator is zero and the slope is undefined, so you should throw an IllegalArgumentException in this case. This will stop your program running and show the specified error message.

Answer

In a Line class, we define two points and a constructor:

```
private Point p1;
private Point p2;
public Line(Point p1, Point p2) {
        this.p1 = p1;
        this.p2 = p2;
}
```

The Point class is designed as:

```java
public class Point {
        private int x;
        private int y;
        public Point() {
        }

        public void setX(int x) {
                this.x = x;
        }

        public int getX() {
                return x;
        }

        public void setY(int y) {
                this.y = y;
        }

        public int getY() {
                return y;
        }
}
```

Now we add a method called getSlope() inside the Line class.

```java
public double getSlope() {
        if (this.p1.getX() == this.p2.getX()) {
                throw new IllegalArgumentException("Denominator
                cannot be 0");
        }
        return (double)(this.p2.getY() - this.p1.getY()) /
        (this.p2.getX() - this.p1.getX());
}
```

The method looks easy, but the tricky part is where we convert the result of division between two integers to a double value, that is,

```
(double)(this.p2.getY() - this.p1.getY()) / (this.p2.getX() -
this.p1.getX())
```

Math: Collinearity

Points are collinear if a straight line can be drawn to connect them. Two basic examples are when three points have the same x- or y-coordinate. The more general case can be determined by calculating the slope of the line between each pair of points and checking whether the slope is the same for all pairs of points.

We use the formula (y2 - y1) / (x2 - x1) to determine the slope between two points (x1, y1) and (x2, y2).

Add the following method to your Line class:

```
public boolean isCollinear(Point p)
```

It needs to return true if the given point is collinear with the points of this line.

CHAPTER 18

Conditional Statements

How do you identify and express the bigger number between the two numbers, x and y?

Math: Hypothesis and Conclusion

In a mathematical formula, we have to introduce the absolute sign to form an expression:

The bigger number between x and y → $\dfrac{(x+y)+|x-y|}{2}$

Recall the if/else structure:

```
if (x >= y) {
        // x is the bigger number
} else {
        // y is the bigger number
}
```

It is very straightforward and fairly easy to read!

The if/else structure follows a common experimentation of hypothesis-to-conclusion.

© Ron Dai 2019
R. Dai, *Learn Java with Math*, https://doi.org/10.1007/978-1-4842-5209-3_18

```
if (<Hypothesis>) {
        <Conclusion>
} else {                    // the hypothesis is NOT valid
        <Different conclusion>
}
```

The <Hypothesis> part needs to be a Boolean value, which may contain one variable or multiple variables in a math expression.

There are several types of structures for the conditional statements (some you've seen, some that will be new to you).

- Simple if, or if/else clause.

- A little more complicated if/else if ladder.

```
if (...) {
      ......
} else if (...) {
      ......
} else if (...) {
      ......
} else {
      ......
}
```

- Nested if/else statement.

```
if (...) {
      ......
} else {
        if (...) {
              ......
        } else {
              ......         ......
        }
}
```

The final pattern is used to implement a tree-like structure. It will depend on the type of problems we solve when we decide which pattern to use.

Example

Is there anything wrong with the following block of code?

```
if (i > 50) {
        <do something...>
} else if (i > 100) {
        <do something...>
} else {
        <do something...>
}
```

Answer

When i <= 50, it will never be i > 100, so the else if branch in the middle is actually a dead path. An easy correction should be to just exchange the position of "50" and "100" in the code. And pay attention to this, when it says else if (i > 50) {...} in the following code block, it actually means 50 < i <= 100.

```
if (i > 100) {
        <do something...>
} else if (i > 50) {
        <do something...>
} else {
        <do something...>
}
```

Example

Create a method to map a student's grades (0 to 100 integers) to a standard GPA score.

Answer

The first solution (v0) uses several if clauses. The problem is, for example, when the marks are 69, it has to execute all four if clauses. This is not an efficient approach.

```java
public static char getGpaScore_v0(int points) {
    if (points > 89) {
        return 'A';
    }
    if (points < 90 && points > 79)        {
        return 'B';
    }
    if (points < 80 && points > 69)        {
        return 'C';
    }
    if (points < 70 && points > 64)        {
        return 'D';
    }
    // if (points < 65) <-- this line can be omitted
    return 'F';
}
```

The second solution (v1) utilizes a nested "if / else" statement. It solves the problem observed from an earlier version - v0.

```java
public static char getGpaScore_v1(int points) {
    if (points > 89) {
        return 'A';
    } else {
        if (points > 79) {
            return 'B';
        } else {
```

```
    if (points > 69) {
            return 'C';
    } else {
            if (points > 64) {
                    return 'D';
            } else {
                    return 'F';
            }
        }
      }
    }
  }
```

To provide better readability into the code structure, the third solution (v2) is introduced as shown.

```
/*
 * 90 to 100 --- A
 * 80 to 89      --- B
 * 70 to 79      --- C
 * 65 to 69  --- D
 * below 65  --- F
 */
public static char getGpaScore_v2(int points) {
        if (points > 89) {
                return 'A';
        } else if (points > 79) {
                return 'B';
        } else if (points > 69) {
                return 'C';
```

```
    } else if (points > 64) {
            return 'D';
    } else {
            return 'F';
    }
}
```

Math: Quadrants

On the Cartesian coordinate system, a quadrant is determined by whether the x and y coordinates are positive or negative numbers. There are four quadrants, separated by the x-axis and the y-axis. Specifically, all the points $(x > 0, y > 0)$ belong to quadrant I (or 1st quadrant); all the points $(x < 0, y > 0)$ belong to quadrant II (or 2nd quadrant); all the points $(x < 0, y < 0)$ belong to quadrant III (or 3rd quadrant); and all the points $(x > 0, y < 0)$ belong to quadrant IV (or 4th quadrant)

Example

Can you write a method to identify which quadrant on the coordinate system that any given point (x, y) belongs to? Both x and y are real numbers. If a point falls on the x-axis or the y-axis, then the method should return 0.

Answer

There are two variables, x and y, in this example. Define x and y as float type of numbers. Do case work analysis as shown below:

Case 1: When a point falls on either x-axis or y-axis ➔ $y = 0$ or $x = 0$
Case 2: When a point falls in the 1st quadrant ➔ $x > 0$ and $y > 0$
Case 3: When a point falls in the 2nd quadrant ➔ $x < 0$ and $y > 0$
Case 4: When a point falls in the 3rd quadrant ➔ $x < 0$ and $y < 0$
Case 5: When a point falls in the 4th quadrant ➔ $x > 0$ and $y < 0$

Combining case 2 and case 5 to a category of x > 0, case 3 and case 4 to a category of x < 0 produces a code structure as:

```
private static int quadrant(float x, float y) {
        if (x == 0 || y == 0) {
                return 0;
        }
        else if (x > 0)                 // x > 0 and y <> 0
        {
                if (y > 0) {
                        return 1;
                }
                return 4;
        }
        else                            // x < 0 and y <> 0
        {
                if (y > 0) {
                        return 2;
                }
                return 3;
        }
}
```

It uses float to hold x and y values, although it could also use double to do so. Both float and double are numeric data types that are used for storing floating-point numbers. The double type requires twice as much space as the float type, as every float type of data is represented in 32 bits while one double type of data uses 64 bits

---

TERNARY OPERATOR

Java enables you to assign a value directly from a Boolean expression (true or false). This is called a ternary operator. For example:

```
int a, b, max;
max = a < b? b : a;
```

This implies, when a < b, max = b; otherwise max = a.

This syntax saves an if/else statement. For example, we may use the following method to get an absolute value:

```
public int getAbsolutionValue(int a) {
        if (a < 0) {
                return -a;
        }
        else {
                Return a;
        }
}
```

But by using the ternary operator, we will get it done by one line of code:

```
a = a < 0 ? -a : a;
```

Problems

1. Please rewrite the code as below to improve its logic and readability (num is an integer value).

    ```
    if (num < 10 && num > 0) {
            System.out.println("It's an one digit
            number");
    }
    ```

```
else if (num < 100 && num > 9) {
        System.out.println("It's a two digit
        number");
}
else if (num < 1000 && num > 99) {
        System.out.println("It's a three digit
        number");
}
else if (num < 10000 && num > 999) {
        System.out.println("It's a four digit
        number");
}
else {
        System.out.println("The number is not
        between 1 & 9999");
}
```

2. Take the following three if statements:

```
if (a == 0 && b == 0) {...}
if (a == 0 && b != 0) {...}
if (a != 0 && b != 0) {...}
```

Please simplify the code logic and combine them together.

CHAPTER 19

Switch Statement

Utilizing a switch conditional statement instead of an if statement can sometimes present clearer code logic. When we have a variable or an expression containing variables that may have different resulting values followed by different actions, it is a good opportunity to use switch.

```
switch (<expression>) {
      case <result 1>:
            <action 1>;
            break;
      case <result 2>:
            <action 2>;
            break;
      ......
      case <result n>:
            <action n>;
            break;
      default:
            <other action>;
            break;
}
```

© Ron Dai 2019
R. Dai, *Learn Java with Math*, https://doi.org/10.1007/978-1-4842-5209-3_19

A simple application of a switch statement is when you take different action under the case of x = 1, or x = 2, or x = 3, ..., as below:

```
switch (x) {
        case 1: ...;
        case 2: ...;
        case 3: ...;
        default: ...;
    }
```

Example

Write a method to print out the month in an English word expression given an integer value input.

Answer

We can use if/else ladder statements to translate an integer to a name of the month.

```
public static void tellNameOfMonthByIfElse(int month) {
        if (month == 1) {
                System.out.println("January");
        } else if (month == 2) {
                System.out.println("February");
        } else if (month == 3) {
                System.out.println("March");
        } else if (month == 4) {
                System.out.println("April");
        } else if (month == 5) {
                System.out.println("May");
        } else if (month == 6) {
                System.out.println("June");
        } else if (month == 7) {
                System.out.println("July");
        } else if (month == 8) {
                System.out.println("August");
```

```java
    } else if (month == 9) {
            System.out.println("September");
    } else if (month == 10) {
            System.out.println("October");
    } else if (month == 11) {
            System.out.println("November");
    } else if (month == 12) {
            System.out.println("December");
    } else {
            System.out.println("Unknown month");
    }
}
```

If we take advantage of the switch conditional statement to do the same translation, it will work well too.

```java
public static void tellNameOfMonthBySwitch(int month) {
        String nameOfMonth;
        switch (month) {
            case 1:   nameOfMonth = "January";
                      break;
            case 2:   nameOfMonth = "February";
                      break;
            case 3:   nameOfMonth = "March";
                      break;
            case 4:   nameOfMonth = "April";
                      break;
            case 5:   nameOfMonth = "May";
                      break;
            case 6:   nameOfMonth = "June";
                      break;
            case 7:   nameOfMonth = "July";
                      break;
```

```java
        case 8:  nameOfMonth = "August";
                 break;
        case 9:  nameOfMonth = "September";
                 break;
        case 10: nameOfMonth = "October";
                 break;
        case 11: nameOfMonth = "November";
                 break;
        case 12: nameOfMonth = "December";
                 break;
        default: nameOfMonth = "Unknown month";
                 break;
    }
    System.out.println(nameOfMonth);
}
```

There is an even better way. How about we define an array to store a list of name strings representing all months in English, that is, the array nameOfMonth. We are essentially building a mapping table between integer numbers (from 0 to 12) and month name strings. Since an array starts with the index 0, we intentionally assign "none" to the first element in the array.

```java
private static String[] nameOfMonth = new String[] {
            "none", "January", "February", "March",
            "April", "May", "June",
            "July", "August", "September", "October",
            "November", "December"
    };

public static void main(String[] args) {
        System.out.println(nameOfMonth[1]);    // January
        System.out.println(nameOfMonth[8]);    // August
    }
```

Example

Write a method to return the number of days in a month, given two integer inputs: year and month.

Answer

Using the switch conditional statement:

```
public static int tellNumberOfDaysByYearMonth(int year,
int month) {
        int numOfDays = 0;
        switch (month) {
            case 1: case 3: case 5:
            case 7: case 8: case 10:
            case 12:
                    numOfDays = 31;
                    break;
            case 4: case 6:
            case 9: case 11:
                    numOfDays = 30;
                    break;
            case 2:
                    if ((year % 4 == 0 && year % 100
                    != 0) || year % 400 == 0) {
                            numOfDays = 29;
                    } else
                            numOfDays = 28;
                    }
                    break;
            default:
                    break;
        }
        return numOfDays;
    }
```

Notice that it has a special logic to handle February for leap years. Because of the complexity in figuring out the total number of days in February, when we apply the static array approach mentioned earlier to this situation, we will have to do the following:

```
private static int[] numberOfDaysByMonth = new int[] {
                0,      // none
                31,     // January
                28,     // February
                31,     // March
                30,     // April
                31,     // May
                30,     // June
                31,     // July
                31,     // August
                30,     // September
                31,     // October
                30,     // November
                31      // December
};
```

Before we use the integer array, we need to modify the value for February according to the year value during runtime:

```
if ((year % 4 == 0 && year % 100 != 0) ||
year % 400 == 0) {
        numberOfDaysByMonth[2] = 29;
}
```

Problem

Use a switch conditional statement to write the following code.

```java
char color ='C';
    if (color=='R') {
            System.out.println("The color is red");
    }
    else if(color=='G') {
    System.out.println("The color is green");
    }
    else if(color=='B') {
            System.out.println("The color is black");
    }
    else {
    System.out.println("Some other color");
    }
```

CHAPTER 20

Tracing Moving Objects

Java provides a basic coding framework, such as `for` or `while` loops and `if` or `switch` conditional statements. We can make use of them to keep track of moving objects versus its times. First, we'll work with a popular math problem - bouncing ball scenario.

Math: Bouncing Ball

In a pure math approach, we'd build a table to record the height after each bounce. It is not that hard, but if we change the height value in the original problem setting, we will have to recalculate the values in the same table by hand.

Example

A ball is dropped from a height of 3 meters. On its first bounce it rises to a height of 2 meters. It keeps falling and bouncing to 2/3 of the height it reached in the previous bounce. On which bounce will it rise to a height less than 0.5 meters? This problem is selected from past AMC 8 (American Mathematics competitions for up to 8th grade)

© Ron Dai 2019
R. Dai, *Learn Java with Math*, https://doi.org/10.1007/978-1-4842-5209-3_20

Answer

The programming approach will largely reduce repetitive manual work.

```java
public static void main(String[] args) {
        System.out.println(ballBouncing(3.0));
}

private static int ballBouncing(double originalHeight) {
        double currentHeight = originalHeight;
        int count = 0;
        while(currentHeight > 0.5) {
                currentHeight = currentHeight * 2 / 3;
                count++;
                System.out.println("Bounce No=" + count +
                                "; current height=" +
                                currentHeight);
        }
        return count;
}
```

When you execute it, the output shows the current height after each bounce.

```
Bounce No=1; current height=2.0
Bounce No=2; current height=1.3333333333333333
Bounce No=3; current height=0.8888888888888888
Bounce No=4; current height=0.5925925925925926
Bounce No=5; current height=0.3950617283950617
5
```

Changing the parameter originalHeight and re-executing the same program will promptly output the detailed result. This is much more efficient than solving it on paper in a traditional mathematical approach.

Example

A snail tries to get out of a well. Each day it climbs up the side of the well 4 feet and each night it slides down the well 2 feet and 6 inches. If the snail starts 40 feet down inside in the morning, how many days will it take the snail take to get out of the well? This problem is selected from a MathIsCool competition.

Answer

In order to keep using an integer value, we convert feet to inches by calculation. Notice that we set the depth of the well as a constant variable by utilizing the keyword `final`. We check if the snail has reached the top of the well, after it climbs up every day, and before it slides down.

```
private static void snail() {
        final int DEPTH = 12 * 40;
        int currentHeight = 0;
        int numOfDays = 0;
        while (currentHeight < DEPTH) {
                currentHeight += 12 * 4;
                numOfDays++;
                if (currentHeight >= DEPTH) {
                        break;
                }
                currentHeight -= 12 * 2 + 6;
                System.out.println("No. " + numOfDays + "
                day - " +
                                (DEPTH - currentHeight) + "
                                inches to the top.");
        }
        System.out.println("No. " + numOfDays + " day -
        at the top.");
}
```

This is partial output from the program runtime:

. . . .

```
No. 1 day - 462 inches to the top.
No. 2 day - 444 inches to the top.
No. 3 day - 426 inches to the top.
No. 4 day - 408 inches to the top.
No. 5 day - 390 inches to the top.
No. 6 day - 372 inches to the top.
No. 7 day - 354 inches to the top.
No. 8 day - 336 inches to the top.
No. 9 day - 318 inches to the top.
No. 10 day - 300 inches to the top.
No. 11 day - 282 inches to the top.
No. 12 day - 264 inches to the top.
No. 13 day - 246 inches to the top.
No. 14 day - 228 inches to the top.
No. 15 day - 210 inches to the top.
No. 16 day - 192 inches to the top.
No. 17 day - 174 inches to the top.
No. 18 day - 156 inches to the top.
No. 19 day - 138 inches to the top.
No. 20 day - 120 inches to the top.
No. 21 day - 102 inches to the top.
No. 22 day - 84 inches to the top.
No. 23 day - 66 inches to the top.
No. 24 day - 48 inches to the top.
No. 25 day - at the top.
```

CHAPTER 21

Counting

We have learned many mathematical methods to solve counting problems. Some of these problems require a deep understanding of permutation and combination. In this chapter, we will learn examples of how to use programming to solve counting problems.

Example

Tickets on a bus were $4.00 and $6.00. A total of 45 tickets were sold and $230 was earned. How many $4.00 tickets were sold? (2007/ MathIsCool problem at http://academicsarecool.com)

Answer

This can be solved by a single loop. We set the variable tickets as the number of $4.00 tickets. Because the total number of tickets is 45, the number of $4.00 tickets cannot be greater than 45. Therefore, tickets is an integer under 46.

```
private static void calculateBusTickets() {
    for(int tickets = 0; tickets < 46; tickets++) {
        int totalMoney = 4 * tickets + 6 * (45 - tickets);
        if (totalMoney == 230) {
            System.out.println(tickets + " $4.00
            tickets were sold.");
            break;
        }
    }
}
```

© Ron Dai 2019
R. Dai, *Learn Java with Math*, https://doi.org/10.1007/978-1-4842-5209-3_21

Example

A chair has 4 legs, a stool has 3 legs, and a table has 1 leg. At a birthday party, there are 4 chairs per table and a total of 18 pieces of furniture. One of the children counts 60 legs total. How many stools are there? (2016/ MathIsCool problem at http://academicsarecool.com)

Answer

In the following method, the variable tables represents number of tables. Then, the number of chairs is 4 × tables, and the number of stools is (18 – tables – 4 * tables).

```
private static void countFurniture() {
        for(int tables = 0; tables < 19; tables++) {
                if (tables + 4 * 4 * tables + 3 * (18 - tables -
                4 * tables) == 60) {
                        System.out.println((18 - tables) + "
                        stools.");
                        break;
                }
        }
}
```

Example

A multiple-choice examination consists of 20 questions. The scoring is +5 for each correct answer, -2 for each incorrect answer, and 0 for each unanswered question. John's score on the examination is 48. What is the maximum number of questions he could have answered correctly? (1987/ AMC8 problem at https://artofproblemsolving.com/wiki/index.php/1987_AJHSME)

Answer

Unlike the previous two examples, in this one we will use two variables, c and w, in the loop. Let's assume the number of correct answers is c, and the number of wrong answers is w. Their sum cannot be greater than

20 – the total number of problems. Because it may have more than one solution, we don't use break to exit the program right after it finds the first solution. This is a different approach from in the previous examples.

```java
private static void scoring() {
    for(int c = 0; c < 20; c++) {
        for(int w = 0; w < 20 - c; w++) {
            if (5 * c - 2 * w == 48) {
                System.out.println("Correct
                answers: " + c
                        + "; wrong answers:
                        " + w);
            }
        }
    }
}
```

The output is:

```
Correct answers: 10; wrong answers: 1
Correct answers: 12; wrong answers: 6
```

Example

How many distinct four-digit numbers are divisible by 3 and have 23 as their last two digits? (2003/10B AMC problem at https://artofproblemsolving.com/wiki/index.php/2003_AMC_8)

Answer

We need to pay attention to the wording, "distinct four-digit numbers," in this problem. The strategy is to separate all conditions into two parts.

- The 4-digit number is divisible by 3 and its last two digits are "23".

- All the four digits are different.

```java
private static void countNumbers() {
        int totalCount = 0;
        for(int i = 1000; i < 10000; i++) {
                if (i % 3 == 0 && i % 100 == 23) {
                        int firstDigit = i / 1000;
                        int secondDigit = i / 100 % 10;
                        if (firstDigit != secondDigit &&
                                firstDigit != 2 &&
                                firstDigit != 3 &&
                                secondDigit != 2 &&
                                secondDigit != 3) {
                                totalCount++;
                                System.out.println(i);
                        }
                }
        }
        System.out.println("Total count = " + totalCount);
}
```

Its output is:

```
1023
1623
1923
4023
4623
4923
5823
6123
6423
6723
7023
```

```
7623
7923
8523
9123
9423
9723
Total count = 17
```

We use one if clause to validate that all four digits are distinct.

```
if (firstDigit != secondDigit &&
       firstDigit != 2 &&
       firstDigit != 3 &&
       secondDigit != 2 &&
       secondDigit != 3) {
```

Alternatively, we may create a general method to check it.

```
if (isDistinct(firstDigit, secondDigit, 2, 3)) {
    ......
}
```

This is the implementation of isDistinct(...).

```
private static boolean isDistinct(int a, int b, int c,
int d) {
      if (a == b) {
            return false;
      } else if (a == c) {
            return false;
      } else if (a == d) {
            return false;
      } else if (b == c) {
            return false;
```

```
        } else if (b == d) {
                return false;
        } else if (c == d) {
                return false;
        } else {
                return true;
        }
    }
```

An improved version in countNumbers2() will be:

```
private static void countNumbers2() {
        int totalCount = 0;
        for(int i = 1000; i < 10000; i++) {
                if (i % 3 == 0 && i % 100 == 23) {
                        int firstDigit = i / 1000;
                        int secondDigit = i / 100 % 10;
                        if (isDistinct(firstDigit, secondDigit,
                        2, 3)) {
                                totalCount++;
                                System.out.println(i);
                        }
                }
        }
        System.out.println("Total count = " + totalCount);
}
```

Example

Ruthie has 10 coins, all either nickels, dimes, or quarters. She has N nickels, D dimes, and Q quarters, where N, D, and Q are all different, and are each at least 1. Amazingly, she would have the same amount of money if she had Q nickels, N dimes, and D quarters. How many cents does Ruthie have? (2012 MathIsCool problem at http://academicsarecool.com)

Answer

Two for-loops are to be used in the following solution.

```
private static void countCoins() {
        for(int n = 1; n < 9; n++) {
                for(int d = 1; d < 10 - n; d++) {
                        int q = 10 - n - d;
                        if (5*n + 10*d + 25*q == 5*q + 10*n + 25*d){
                                System.out.println((5*n+10*d+25*q)+
                                " cents.");
                                System.out.println("N="+n+";
                                D="+d+"; Q="+q);
                        }
                }
        }
}
```

Output is:

```
155 cents.
N=1; D=5; Q=4
```

Example

Three friends have a total of six identical pencils, and each one has at least one pencil. In how many ways can this happen? (2004 AMC8 problem at https://artofproblemsolving.com/wiki/index.php/2004_AMC_8)

Answer

We use two for-loops to simulate how we distribute the six identical pencils to three people represented by variables, first, second, third.

```
private static void countWays() {
        int count = 0;
        for(int first=0; first <= 6; first++) {
```

```
            for(int second = 0; second <= 6 - first;
            second++) {
                    int third = 6 - first - second;
                    if (first > 0 && second > 0 && third > 0) {
                            count++;
                                    System.out.println("first="
                                    + first + "; second=" +
                                    second + "; third=" +
                                    third);                              }
                }
        }
        System.out.println("Total count=" + count);
}
```

The output is:

```
first=1; second=1; third=4
first=1; second=2; third=3
first=1; second=3; third=2
first=1; second=4; third=1
first=2; second=1; third=3
first=2; second=2; third=2
first=2; second=3; third=1
first=3; second=1; third=2
first=3; second=2; third=1
first=4; second=1; third=1
Total count=10
```

Example

Seven distinct pieces of candy are to be distributed among three bags.
The red bag and the blue bag must each receive at least one piece of candy;
the white bag may remain empty. How many arrangements are possible?
(2010/10B AMC problem at https://artofproblemsolving.com/wiki/
index.php/2010_AMC_10B)

Answer

First, we design an experiment. In this experiment, we want to put seven distinct strings ("A," "B," "C," "D," "E," "F," "G") into three String arrays. The seven strings represent seven distinct pieces of candy. The three arrays represent red, blue, and white bags. The order of the placement doesn't matter, but we need to make sure only the last array can be empty after the placement is made. The goal is to find the number of different placements.

When we place "A" in one of the three arrays, we need a for-loop for three different arrays. Then we need another for-loop to place "B," and so on; in total we will need seven for-loops. In each for-loop, we append the string to the existing array, that is, bag[i]. But after it is done, we need to remove it from the tail of the array string in order for it to try the next option. This is why we use bag[i].replace("A", ""). The same applies to the other six strings.

We come up with a "straightforward" but ugly-looking version as what follows.

Note   BAG – red: 0, blue: 1, white: 2; 3 in the loops represent the three string arrays, that is, the three bags.

```
/// BAG - red: 0, blue: 1, white: 2
private static void distributeCandy() {
    int count = 0;
    String[] bag = { "", "", "" };
    for(int i=0; i < 3; i++) {
        bag[i] += "A";
        for(int j=0; j < 3; j++) {
            bag[j] += "B";
            for(int k=0; k < 3; k++) {
```

```
                bag[k] += "C";
                for(int l=0; l < 3; l++) {
                        bag[l] += "D";
            for(int m=0; m < 3; m++) {
                bag[m] += "E";
                for(int n=0; n < 3; n++) {
                    bag[n] += "F";
                    for(int p=0; p < 3; p++) {
                        bag[p] += "G";
                        if(bag[0].length() > 0 && bag[1].length()
                        > 0) {
                                    count++;
                            System.out.println("Red=" + bag[0] + "
                            Blue=" + bag[1] + " White=" + bag[2]);
                        }
                        bag[p] = bag[p].replace("G", "");    }
                bag[n] = bag[n].replace("F", "");    }
            bag[m] = bag[m].replace("E", "");    }
                bag[l] = bag[l].replace("D", "");    }
            bag[k] = bag[k].replace("C", "");    }
        bag[j] = bag[j].replace("B", "");    }
    bag[i] = bag[i].replace("A", "");    }
    System.out.println("Total count: " + count);
}
```

The code structure looks too complicated. There are too many nested for-loops. A better idea is to apply a recursive approach to improve its simplicity. Now see a new version:

```
private static int count = 0;
private static String[] bag = { "", "", "" };
private static String[] CANDY = new String[] { "A", "B", "C",
"D", "E", "F", "G" };
```

```java
private static String RemoveLastChar(String s) {
    if (s == null && s.length() < 1) {
        System.out.println("Input string is invalid!");
        return "";
    }
    return s.substring(0, s.length() - 1);
}
private static void distributeCandies_Recursive(int pointer) {
    for(int i=0; i < 3; i++) {
        bag[i] += CANDY[pointer];
        if (pointer == CANDY.length - 1) {
            if (bag[0].length() > 0 && bag[1].
            length() > 0) {
                count++;
                System.out.println("Red=" + bag[0]
                + " Blue=" + bag[1] + " White=" +
                bag[2]);
            }
        }
        else {
            distributeCandies_Recursive((pointer + 1));
        }
        bag[i] = RemoveLastChar(bag[i]);
    }
}
```

We then include two following lines in the main function for execution.

```java
distributeCandies_Recursive(0);
System.out.println("Total count: " + count);
```

Example

A palindrome between 1000 and 10000 is chosen at random. What is the probability that it is divisible by 7? (2010/10B AMC problem at https://artofproblemsolving.com/wiki/index.php/2010_AMC_10B)

Answer

A palindrome number is a number that reads the same from its left to right as from its right to left.

In the isPalindrome() method, we reverse the number string and compare it with the original number string. If the reversed string turns out to be the same as the original one, it is identified as a palindrome string. For "8558." its reversed string "8558" is the same as itself.

We introduce StringBuffer class to leverage its reverse() method. Every number within the range [1000, 10000] is converted to a string type, before it is passed to the isPalindrome() method. The solution can be applied to any range of integer numbers.

```
public static void main(String[] args) {
        countDivisibility();
}

private static boolean isPalindrome(String numberStr) {
        String reversed = new StringBuffer(numberStr).reverse().
        toString();
        return reversed.equals(numberStr);
}

private static void countDivisibility() {
        int count = 0;
        int total = 0;
        for(int i = 1000; i < 10001; i++) {
                if(isPalindrome(Integer.toString(i))) {
                        total++;
                        if (i % 7 == 0) {
                                count++;
                                System.out.println(i);
```

```
            }
        }
    }
    System.out.println("Probability=" + count + "/" + total);
}
```

There is a different way in the method isPalindrome2(), which contains the same functionality as what the isPalindrome() method has. Instead of using StringBuffer, you may simply compare each character from the leading half of one with the trailing half of the original string.

```
private static boolean isPalindrome2(String s) {
    int len = s.length();
    for( int i = 0; i < len / 2; i++ ) {
                if (s.charAt(i) != s.charAt(len - i - 1)) {
                    return false;
                }
    }
    return true;
}
```

Example

A base-10 three-digit number n is selected at random. What is the probability that the base-9 representation and the base-11 representation of n are both three-digit numbers? (2003/10A AMC problem at https://artofproblemsolving.com/wiki/index.php/2003_AMC_10A_Problems)

Answer

The total count of base-10 three-digit numbers is 900. A three-digit number 124 in base-10 is 147 in base-9, and 103 in base-11; 720 in base-10, 880 in base-9, and 5A5 in base-11.

The crucial method is countBase10Numbers():

```
private static void countBase10Numbers() {
    int count = 0;
    for(int i = 100; i < 1000; i++) {
```

```
            String base9Number = convertToBaseN(i, 9);
            String base11Number = convertToBaseN(i, 11);
            if (base9Number.length() == 3 &&
                    base11Number.length() == 3) {
                count++;
                System.out.println(i + " -> " +
                base9Number +
                                "; " + base11Number);
            }
        }
        System.out.println(count + " out of " + (1000 - 100));
}
```

The supporting method is convertToBaseN().

```
private static String convertToBaseN(int base10, int n) {
        if(n < 2 || n > 16) {
                return "";
        }
        String baseN = myOneDigit[base10 % n];
        base10 = base10 / n;
        while(base10 > 0) {
                baseN = myOneDigit[base10 % n] + baseN;
                base10 = base10 / n;
        }
        return baseN;
}
```

The myOneDigit array is:

```
private static String[] myOneDigit =
{ "0", "1", "2", "3", "4", "5", "6", "7", "8", "9", "A", "B",
"C", "D", "E", "F" };
```

It is actually equivalent to a method with implementation of the switch conditional statement:

```
private static String convertDigit(int digit) {
        String s = "";
        switch(digit) {
                case 10:
                        s = "A";
                        break;
                case 11:
                        s = "B";
                        break;
                case 12:
                        s = "C";
                        break;
                case 13:
                        s = "D";
                        break;
                case 14:
                        s = "E";
                        break;
                case 15:
                        s = "F";
                        break;
                default:
                        s = Integer.toString(digit);
                        break;
        }
        return s;
}
```

Obviously, the approach using a string array is simpler.

Problems

1. Richa and Yashvi are going to Jamaica with their school. They plan on attending a fair where the admission for children is $1.50 and $4.00 for adults. On a specific day, 2,200 people enter the fair and $5,050 is collected. How many children attended? (2017 MathIsCool)

2. In a mathematics contest with 10 problems, a student gains 5 points for a correct answer and loses 2 points for an incorrect answer. If Olivia answered every problem and her score was 29, how many correct answers did she have? (2002 AMC8)

3. How many positive integers not exceeding 2001 are multiples of 3 or 4 but not 5? (2001 AMC10)

4. How many positive three-digit numbers contain exactly two distinct digits (e.g., 343 or 772, but not 589 or 111)? (2006 MathIsCool)

5. Rebecca goes to the store where she buys five plants. If the store sells three types of plants, how many different combinations of plants can she buy? (2005 MathIsCool)

CHAPTER 22

Factorization

In school math, we usually follow a procedure to find all factors for any given positive integer number. This process is called factorization, and it takes quite a lot of calculation depending on how big the integer number is. With the Java programming environment, let's create a simple program to do the same job for us. We'd like to write code to find all factors of any positive integer. When a user inputs "10," the program is supposed to output: 1, 2, 5, and 10. As the first step, we need to define the procedure and then implement it by Java code.

Math: Finding Factors

Recalling how we found factors manually in school, we used an integer number from the smallest (i.e., "1") up to the largest (i.e., the given integer itself), one by one, to check if it is divisible by the given number. When the answer was yes, we knew it was a factor. Otherwise, we skipped it and moved to the next number.

We create the following block of code to accomplish the procedure. We label this version of code as "v1," and we plan to make improvements from here.

```java
private static int listFactors_v1(int n) {
        int counter = 0;
        for (int i = 1; i <= n; i++) {
                if (n % i == 0) {
```

```
                    if (counter > 0) {
                            System.out.print(", ");
                    }
                    System.out.print(i);
                    counter++;
            }
        }
        System.out.println();
        System.out.println("Number of factors: " + counter);
        return counter;
}
```

The main method will look like this:

```
public static void main(String[] args) {
        Scanner input = new Scanner(System.in);
        int iterations = 0;
        while (true) {
                iterations++;
                System.out.println("Enter an integer number:");
                int k = input.nextInt();
                if (k < 0) {
                        k = -k;
                }
        }
        System.out.println("Number of factors: " +
        listFactors_v1(k));
        input.close();
}
```

When you compile and execute the code, you will see an output as:

```
Enter an integer number:
2018
1, 2, 1009, 2018
Number of factors: 4
```

Don't think that we have a perfect solution to the problem. Actually, it is far from complete.

Math: Halving the Problem

When we iterate every single number from 1 to n, in order to find all possible divisors of n, we observe that any integer greater than n/2 will not be a divisor of n. Therefore, we only need to check from 1 up to n/2, instead of n. This change will save half of the iterations in the program. So, we now have an immediate improvement in the following version 2.1.

```java
private static int listFactors_v21(int n) {
        int counter = 0;
        for (int i = 1; i <= n / 2; i++) {
                if (n % i == 0) {
                        if (counter > 0) {
                                System.out.print(", ");
                        }
                        System.out.print(i);
                        counter++;
                }
        }
        System.out.println(", " + n);
        counter++;
        System.out.println("Number of factors: " + counter);
        return counter;
}
```

In addition to the algorithm change, we will remove one `if` clause to reduce the complexity by a little bit. Then we will come up with the following version 2.2 with minor modifications.

```java
private static int listFactors_v22(int n) {
        System.out.print("1"); // "1" is always the 1st factor
        int counter = 1;
        for (int i = 2; i <= n / 2; i++) {
                if (n % i == 0) {
                        System.out.print(", " + i);
                        counter++;
                }
        }
        System.out.println(", " + n); // n is always the last
        factor
        counter++;
        System.out.println("Number of factors: " + counter);
        return counter;
}
```

Is it good enough? Actually not.

Math: Using the Square Root

If we remember in math how to test whether a positive integer number is prime or not, we only use integers from 2, 3, up to the square root of n. We will apply the same logic here to find a pair of factors in order to reduce the number of iterations.

```java
private static int listFactors_v31(int n) {
        int counter = 0;
        for (int i = 1; i <= Math.sqrt(n); i++) {
```

```
        if (n % i == 0) {
                if (counter > 0) {
                        System.out.print(", ");
                }
                System.out.print(i);
                counter++;
                if (i != n / i) {
                        System.out.print(", " + n / i);
                        counter++;
                }
        }
    }
    System.out.println();
    System.out.println("Number of factors: " + counter);
    return counter;
}
```

The current version 3.1 is obviously better, because it only checks numbers up to the square root of n, instead of n/2. Using n=100 as an example, now we check from 1 to 10, not from 1 to 50. Obviously, the reduction of iterations is significant. But we quickly discover an issue with it. The output from the current solution is:

```
Output:

        Enter an integer number:
        2018

        1, 2018, 2, 1009
        Number of factors: 4
```

The list of factors is not in ascending order as we hoped, simply because it prints out factors by pairs. To resolve the issue, we will create two strings. One string stores the smaller one from every pair of factors. The other string stores the bigger one from each pair.

```
private static int listFactors_v32(int n) {
        String s1 = "1";
        String s2 = Integer.toString(n);
        int counter = 2;
        for (int i = 2; i <= Math.sqrt(n); i++) {
                if (n % i == 0) {
                        s1 += ", " + i;
                        counter++;
                        if (i != n / i) {
                                s2 = n / i + ", " + s2;
                                counter++;
                        }
                }
        }
        System.out.println(s1 + ", " + s2);
        System.out.println("Number of factors: " + counter);
        return counter;
}
```

Can we possibly make further improvements from the current version 3.2? The answer is still YES. Instead of storing the smaller number from each pair of divisors to a string, we send it directly to the console. This will save memory space of one string. It is another "little" change, but potentially a big save, when we deal with a big number that may have a huge list of factors. We finally landed on v3.3:

```
private static int listFactors_v33(int n) {
        String s = Integer.toString(n);
        int counter = 2;
        System.out.print("1");
        for (int i = 2; i <= Math.sqrt(n); i++) {
                if (n % i == 0) {
```

```
                System.out.print(", " + i);
                counter++;
                if (i != n / i) {
                        s = n / i + ", " + s;
                        counter++;
                }
            }
        }
    System.out.println(", " + s);
    System.out.println("Number of factors: " + counter);
    return counter;
}
```

Here is a summary of what we have done to solve this problem:

- We kept thinking about how to improve our implementation according to three basic rules:

  (a) Adopt the best algorithm we know;

  (b) Optimize code to consume less memory and run faster;

  (c) Write code that is easy to understand for future maintenance.

- We reduced the actual upper bound of the integer number from n to n/2, then to the square root of n.

- We avoided having to use an extra string for temporary storage.

All these efforts have contributed to a well-optimized code. This is indeed the art of programming.

CHAPTER 23

Exploratory Experimentation of Pi

Scientists must always keep track of the population of fish in order to monitor the impact on the fish life cycle from natural environmental changes. There is a type of fish called AA in a lake for research. The scientists set free a small group of labeled fish AA, whose total number equals to #(total_labeled), to the lake. After a period of time, they capture a number of fish AA randomly from the lake and use them as samples, whose total number equals to #(total_captured). Among these sample fish AA, they sort out the labeled fish AA, whose total number equals to #(labeled_among_captured).

Math: Calculating a Population

Assuming all the fish, including the labeled and the unlabeled, are distributed evenly in the lake, we use the following formula to figure out the current population of fish AA in the lake.

```
#(total_captured) * #(total_labeled) / #(labeled_among_captured)
```

The formula is expressed in a simple ratio form. The more evenly distributed fish AA is in the lake, the more accurate result the formula will produce. It is essentially a statistical idea that uses a small pool of sample

© Ron Dai 2019
R. Dai, *Learn Java with Math*, https://doi.org/10.1007/978-1-4842-5209-3_23

data to predict a possibly large total number in a big picture. It attempts to make a measurement from the unmeasurable object with a minimized error margin.

The nature of the fish experimentation is based on probability theory. It also applies to many other interesting problem areas. One of them is to compute Pi: 3.14159.........

Example

How can we do basic programming to calculate the value of Pi?

Math: Pi from Probability Theory

We inscribe a circle into a square that leans closely against the x-axis and y-axis in a Cartesian coordinate plane. Say the length of the square, which is the diameter of the circle, is n. The area of the circle can be presented in a formula as:

$$p \times (n/2)^2$$

p is something to figure out, that is, Pi.

If one randomly selects a point within the square, what is the probability that the point is exactly inside the circle?

We know the answer after having learned the basic geometric probability. It should be the ratio of an area between the circle and the square. Since the area of the square is n^2, the ratio will be p/4.

$$p \times (n/2)^2 : n^2 = p : 4$$

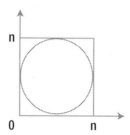

Answer

The probability p/4 indicates that if we repeat the same point selection process as many times as possible, the ratio of the number of points selected inside the circle versus the number of points selected inside the square will be approaching to (and eventually equal to) p/4. Therefore, once we find out the ratio, we know an approximation of Pi. This is the algorithm we will use in the program.

```
public static double computePi(int total, int n) {
        int count = 0;
        for(int i=0; i < total; i++) {
                double x = n * Math.random();
                double y = n * Math.random();
                if ((x - n/2) * (x - n/2) + (y - n/2) * (y - n/2)
                                < (n/2) * (n/2)) {
                        count++;
                }
        }
        return (double)count * 4 / total;
}
```

In this method,

- The integer parameter value `total` is the total number of experiments (i.e., total number of the selected sample points).

- The integer parameter value n is the side length of the square, or the diameter of the circle.

- `Math.random()` is a Java built-in function from `java.util.Random` package. It generates a random number in double between 0 and 1. Multiplying it with n makes the x and y coordinates' values between 0 and n.

- The inequality "$(x - n/2)^2 + (y - n/2)^2 < (n/2)^2$" is to check if the point lies inside the circle, whose center point is at $(n/2, n/2)$.

You may call the method by following lines from the main method():

```java
public static void main(String[] args) {
        int onehMillion = 100 * 1000 * 1000;
        for(int i=0; i < 10; i++) {
                System.out.println(computePi(onehMillion,
                100));
        }
}
```

- n = 100 is the side length. It doesn't contribute to the formula directly. You may use other numbers like 10 or 2 or any even numbers (due to the "n/2") for an extended experiment purpose.

- `int onehMillion = 100 * 1000 * 1000` equals to 100 million. It indicates the total number of points we pick in one exploratory test. The `100 * 1000 * 1000` is a multiplication operation that will be executed during the compilation time, prior to runtime. There is no implication of extra computing time by the current multiplicative expression in coding.

- The `for-loop()` drives the same experiments by multiple times, that is, 10 times.

The output on the console will be like this:

```
3.14150492
3.14166436
3.14157872
3.14143904
3.14174756
3.14153872
3.14161072
3.14155196
3.14198448
3.14158056
```

With less than 10 lines of code in the method, it takes about 5 seconds to complete the execution and output the estimated Pi value.

Last but not least, we will need to change `int` to `long` if the values of `total` and `i` are too big. Remember that `int` type of data can be up to 32 bits, which equals to 2^{32}. When the total is a larger number than that, we will need to use `long` type, which supports 64 bits (equal to 2^{64}).

```
public static double computePi(long total, int n) {
        long count = 0;
        for(long i=0; i < total; i++) {
                double x = n * Math.random();
```

```
        double y = n * Math.random();
        if ((x - n/2) * (x - n/2) + (y - n/2) * (y - n/2)
                    < (n/2) * (n/2)) {
            count++;
        }
    }
    return (double)count * 4 / total;
}
```

The long value type of parameter needs to be passed as . . . L as shown next. However, it will take much longer to execute, unless it runs on a high-computing power PC.

```
public static void main(String[] args) {
    long hugeNumber = 1000 * 1000 * 1000 * 1000L;
    for(int i=0; i < 10; i++) {
        System.out.println(computePi(hugeNumber, 100));
    }
}
```

If we increase the value of total, it will have more coverage of the area by a larger number of points and return a more accurate result of Pi. For the 100-million points we pick in this program, it almost guarantees to find out Pi = 3.141..., at the thousandth precision level. To achieve more precision, we will need a more powerful computing machine. At least we know that with an ideal computing platform, we will be able to nail down the Pi value at a designated precision level.

From this example, we have learned that we can apply the same ratio and probability concepts to finding out an area surrounded by any curve, as long as we know the functional model of the curve.

Example

With the probability concept in mind, use Java programming to find out the area among line x = 0, y = 0, and curve y = -2x² + 12x -18.

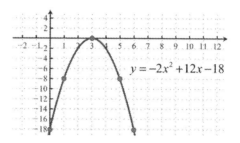

Answer

This approach has helped us solve a problem, while a pure mathematical solution originally requires knowledge of Calculus.

```
public static double computeArea(int total) {
        int count = 0;
        for(int i=0; i < total; i++) {
                double x = Math.random() * 3;
                double y = Math.random() * -18;
                if (-2 * x * x + 12 * x - 18 < y) {
                        count++;
                }
        }
        return (double)count * 54 / total;
}
```

Problem

Create a program to find out Euler's number e.

CHAPTER 24

Classes in Object-Oriented Programming

An object is essentially a representation of a thing. An object has some attributes, just like everything has characteristics. However, a class in the programming world is primarily a data structure designed for a specific object. The class is also said to be a blueprint of an object. It keeps track of a bundle of related things about the object. These related things are known as fields, properties, and functions (or methods).

The fields are data members of a class. They must be declared and initialized before they are used. They are mostly for class internal use.

Some fields may serve as properties, which are attributes of an object (e.g., an employee's name or a bank account's balance).

The properties can be changed by setters, and they can be accessed by getters from outside the class.

Getters and setters are methods to hide the internal implementations of the class properties. This design enables developers to update some of the existing implementations easily later on. It is an example of encapsulation characteristics of object-oriented programming.

© Ron Dai 2019
R. Dai, *Learn Java with Math*, https://doi.org/10.1007/978-1-4842-5209-3_24

In the following example class `Student`,

- `firstName, lastName, age` are all fields. Since all these fields have getter/setters, for example, `getAge()`, `setAge()`, they are also properties.

- `public Student()` is the default constructor defined in the `Student` class. It is like a method, and it is executed when the object is created from the `Student` class, that is:

  `Student student = new Student();`

- `public Student(String firstName, String lastName)` is another constructor of the `Student` class. It instantiates (or creates) an object by directly assigning `firstName` and `lastName` to the fields, that is:

  `Student student = new Student("John", "Doe");`

- `public String getFirstName()`, `public String getLastName()`, and `public int getAge()` are methods to query values of the private fields `firstName, lastName, age`. They are getters.

- `public void setAge(int age)` is a method to assign a value to the private field `age`. It is a setter.

- The keywords `public` and `private` in front of data types (e.g., `String, int`), or method names, are called access modifiers. The `public` means it is visible to all, while the `private` is only open to the current class scope.

```
public class Student {
        private String firstName;
        private String lastName;
        private int age;
```

```
public Student() {
}

public Student(String firstName, String
lastName) {
        this.firstName = firstName;
        this.lastName = lastName;
}

public String getFirstName() {
        return this.firstName;
}

public String getLastName() {
        return this.lastName;
}

public int getAge() {
        return this.age;
}
public void setAge(int age) {
        this.age = age;
}
}
```

A simple structural view is:

```
class Student {
        //fields, e.g. firstName, lastName, and age

        //constructor

        //setter/getter methods, e.g. get/set firstName,
        lastName, and age

}
```

The keyword this inside the Student class is a reference to the current object, whose fields (e.g., firstName, lastName, age) are being used. By using this, the current object's methods or constructors can be invoked as well.

There are two kinds of so-called non-fields in a program.

- local variables inside methods

- parameters x1 and x2 in a method like: myMethod (x1, x2)

Relationship between class and object:

- Can I have an object without having a class? NO

- Can I have a class without having an object? YES

- Can I create multiple instances of a class? YES

Lab Work

Create a class called Name that represents a person's name. The class should have fields named firstName representing the person's first name, lastName representing their last name, and middleInitial representing their middle initial (a single character). Your class should contain only fields for now.

Lab Work

Create the outline of a public class named Vehicle.

```
public class Vehicle {
    ......
}
```

And then create a main program to operate the Vehicle object created.

Lab Work

Add a constructor to the Point class shown next that accepts another Point as a parameter and initializes the new Point to have the same (x, y) values. Use the keyword this in your solution.

Then, add methods named setX and setY to the Point class. Each method accepts an integer parameter and changes the Point object's x- or y-coordinate to be the value passed, respectively.

```
public class Point {
 int x;
 int y;
 // your code goes here
}
```

Problems

1. How is an object different from a class?

    a) Objects are used in object-oriented programming and classes are used in class-oriented programming.

    b) An object is an entity that encapsulates related data and behavior, while a class is the blueprint for a type of object.

    c) An object is not encapsulated, and a class is encapsulated, making classes more powerful and reusable than objects.

    d) An object is a kind of class that does not contain any behavior.

    e) A class is an instance of an object. One object can be used to create many classes.

2. Describe a real-world scenario about how the concepts of class and object are being used.

3. Design and write a class called Game. You will need to think about what kind of fields this class should have, and then add several methods to enrich the class.

    For example,

    (1) you may define the price of the game;

    (2) you may classify the game as "computer game" or "video game";

    (3) you may define the platform on which the game can be used, such as "xbox," "playstation," "nintendo," etc.;

    (4) you may define a constructor with a parameter;

    (5) you may define methods to set/get any fields as mentioned above;

    (6) anything you can think about a Game, when you want to define it as a class.

4. Given the following class, called NumberHolder, write some code that creates an instance of the class, initializes its two member variables, and then displays the value of each member variable.

```
public class NumberHolder {
 public int anInt;
 public float aFloat;
}
```

5. Which of the following are differences between a
    field and a parameter? There might be multiple
    answers to this question.

    (A) A field is a variable that exists inside of an object,
        while a parameter is a variable inside a method
        whose value is passed in from outside.

    (B) Fields can store many values while parameters
        can store only a single value. Field syntax differs
        because they can be declared with the `private`
        keyword.

    (C) Parameters must be primitive types of values,
        while fields can be objects.

    (D) A field's scope is throughout the class, while a
        parameter's scope is limited to the method.

    (E) A field takes up more memory in the computer
        than a parameter does.

    (F) You can only have one field per class, while you
        can have as many parameters as you want.

    (G) Fields are constant and can be set only once,
        while parameters change on each call.

6. Suppose a method in the `Account` class is defined as:

    ```
    public double computeInterest(int rate)
    ```

    And suppose the client code has declared an
    `Account` variable named `acct`. Which of the
    following would be a valid call to the above method?

(A) `int result = Account.computeInterest(14);`

(B) `double result = acct.computeInterest(14);`

(C) `double result = computeInterest(acct, 14);`

(D) `new Account(14).computeInterest();`

(E) `acct.computeInterest(14, 15);`

CHAPTER 25

Interface – Total Abstraction

The concept of interface is a part of abstraction, one of the four OOP. characteristics. Abstraction is about an abstract design of common features, including operations of the object.

Interface is the blueprint of a class. However, it is neither a class nor an object. All methods defined in an interface are abstract. There are no implementation details allowed inside any method of the interface. The class that is going to implement the interface will take care of the actual implementation of the methods.

Let's look at an example of an interface and classes implemented from it. Auto is a general term representing vehicles. We use Auto to define an interface.

```java
public interface Auto {
        void start();
        void stop();
        void turn();
        void back();
        void park();
}
```

© Ron Dai 2019
R. Dai, *Learn Java with Math*, https://doi.org/10.1007/978-1-4842-5209-3_25

As two common types of automobiles, cars and buses are common objects. Both cars and buses share the same type of behaviors defined in Auto interface. When we create a class for cars and buses, we use the keyword implements to implement car and bus from the same Auto interface with different behavioral details.

We use class Car as an example:

```
public class Car implements Auto {
      private String maker;

      public void start() {
            // car starts its engine
      }

      public void stop() {
            // car stops its engine
      }

      public void turn() {
            // car turns left or right at a corner
      }

      public void back() {
            // car backs
      }

      public void park() {
            // car parks
      }

      public String getMaker() {
            return this.maker;
      }
```

```
public void setMaker(String maker) {
        this.maker = maker;
    }
}
```

As you probably have noticed,

- An interface indicates what the object can do.

- When a class implements the interface, it defines what the object is doing with the necessary details.

The design of the interface allows developers to modify the underlying classes without altering the callers' implementations; this is sometimes called coding to the interface. There are at least two circumstances when we should consider adopting an interface design.

- When we want to only specify the behavior of a particular data type, without being concerned about whoever implements its behavior.

    For example:

    We define an interface Auto that is an abstract concept and a general term. In this interface, we define several methods such as start, stop, turn, back, and park by their signatures, without any implementation details. We will add implement details in these methods when we create Car or Truck classes that implement the Auto interface.

– When all classes have the same structure, but they
totally have different functionalities.

For example:

Dogs and cats communicate in totally different
ways. A dog barks, but a cat meows. We may define
an interface Animal and create a class Dog and class
Cat, like shown here.

```java
public interface Animal {
        public void communicate();
}

public class Dog implements Animal {
        public void communicate() {
                System.out.println("bark, bark!");
        }
}

public class Cat implements Animal {
        public void communicate() {
                System.out.println("meow, meow...");
        }
}
```

Java supports multiple interface implementation: for example, if we
define another interface, MovingObject as shown.

```java
public interface MovingObject {
        void movingNorth();
        void movingSouth();
}
```

Class Car can implement from both interfaces, Auto and MovingObject as shown here.

```
public class Car implements Auto, MovingObject {
        ...
        public void movingNorth() {
                // car moves North
        }

        public void movingSouth() {
                // car moves South
        }
}
```

CHAPTER 26

Inheritance – Code Reuse

As one of the OOP principles, inheritance is designed to centralize the common functionality of many different objects. As a result of that, it reduces duplicated code in many classes.

Inheritance introduces two types of classes: "superclass" and "subclass." The subclass inherits from the superclass. The superclass is the same thing as the "base class." The subclass contains not only all the methods and the fields inherited from the superclass, but also other methods and fields defined by the subclass.

For example, we define a new class called Sedan that inherits from the Car class we created earlier. The Car class implements an interface called Auto. In the Sedan class, we define a Boolean field isFourDoorHatchback and a method called isFourWheelDrive().

The keyword extends is used to describe the class that Sedan inherits from class Car.

```java
public class Sedan extends Car {
        public Boolean isFourDoorHatchback;
        public Boolean isFourWheelDrive(){
                return true;
        }
}
```

© Ron Dai 2019
R. Dai, *Learn Java with Math*, https://doi.org/10.1007/978-1-4842-5209-3_26

We then create a `main` method in a `Driver` class to play with the `Sedan` class.

```
public class Driver {
        public static void main(String[] args) {
                Sedan sedan = new Sedan();
                sedan.start();
                sedan.stop();
                sedan.turn();
                sedan.back();
                sedan.park();
                sedan.setMaker("Toyota");
                sedan.getMaker();

                sedan.isFourDoorHatchback = true;
                sedan.isFourWheelDrive();
        }
}
```

As you see, the `Sedan` object (i.e., sedan) has all the methods and fields inherited from its superclass `Car`. In addition to that, `Sedan` class has its own method and field. In the same `main` method, we add more code:

```
                Car car = new Sedan();
                car.start();
                car.stop();
                car.turn();
                car.back();
                car.park();
                car.setMaker("Toyota");
                car.getMaker();
```

This example tells us that we can create an object from a superclass (i.e., Car) instantiated from its subclass (i.e., Sedan). All the methods and fields under its superclass are available as expected, but the methods and fields under its subclass are not accessible.

If we try to do:

```
Sedan sedan2 = new Car();
```

We will get error message:

`"Type mismatch: cannot convert from Car to Sedan".`

This clearly tells us that we are not allowed to create a subclass object (i.e., Sedan) instantiated from its superclass (i.e., Car).

In Java, however, it doesn't support multiple inheritance. Instead, it uses an interface to achieve the same goal as what multiple inheritance attempts to do in other programming languages.

Problems

1. Which of the following is the correct syntax to indicate that class A is a subclass of B?

   (a) `public class A : super B {`

   (b) `public class B extends A {`

   (c) `public class A extends B {`

   (d) `public A(super B) {`

   (e) `public A implements B {`

2.   Consider the following classes:

public class Vehicle {...}

public class Car extends Vehicle {...}

public class SUV extends Car {...}

Which of the following are legal statements?

(a) `Car c = new Vehicle();`

(b) `SUV s = new SUV();`

(c) `SUV s = new Car();`

(d) `Car c = new SUV();`

(e) `Vehicle v = new Car();`

(f) `Vehicle v = new SUV();`

CHAPTER 27

Encapsulation and Polymorphism

In addition to "abstraction" and "inheritance," there are another two principles in OOP, "encapsulation" and "polymorphism."

Encapsulation

You may have heard the phrase "information hiding," which intends to conceal the detailed implementations of an object behind a higher level of abstraction. Information hiding is mainly for security concerns, while encapsulation is to keep data and class implementation details inside a class for complexity concerns. However, encapsulation combines internal data and methods and enables its internal data to be accessible from outside through its public methods. And the class has private instance variables that are only accessible by methods in the same class. This helps managing code that is to be updated frequently. This is known as: "encapsulate what varies," which is one of the best practice design principles.

© Ron Dai 2019
R. Dai, *Learn Java with Math*, https://doi.org/10.1007/978-1-4842-5209-3_27

In the Student class created in an earlier chapter, we have the following private field and public methods.

> `private int age;` ← only accessible from inside class
>
> `public void setAge(int age);` ← setter accessible from outside
>
> `public int getAge();` ← getter accessible from outside

This is a simple example of encapsulation, in terms of how we set a student's age value and how we access the age information.

```java
public class TestStudent {
    public static void main(String[] args) {
        Student student = new Student("John",  "Doe");
        /*
            student.age = 20;
            This line will give compiler error
            age field can't be used directly as it is
            private
        */
        student.setAge(20);
        System.out.println("Student name: " + student.
getFirstName() + " " + student.getLastName() + ";
age: " + student.getAge());
    }
}
```

There is a difference between abstraction and encapsulation. Abstraction is hiding complexity (i.e., implementation details) by using interfaces, while encapsulation is wrapping code (i.e., implementation) and data (i.e., value of variables) within the same class.

Polymorphism

"Poly" means many. "Morph" indicates form or shape. "Polymorphism" is an object's ability to present the same interface with many different forms. There are many examples of this in Java programming design.

- With one interface, we can create multiple classes. Each class implements the same method with different details.

- In the basic class design, we can create multiple constructors with different input parameters.

- Similarly, we can use the same method name with a different set of input parameters in a class design. This is also called "overloaded methods."

- In a subclass, we can "override a method" defined originally in its superclass.

Problems

1. Write an interface called GeometricObject, which declares two abstract methods: getPerimeter() and getArea().

2. Write the implementation class Circle, with a protected variable radius, which implements the interface GeometricObject.

CHAPTER 28

Array – a Simple and Efficient Data Structure

When we are in a situation where we need to store and manipulate a bunch of the same types of data, we need to think about the right data structure to use. Let's say we want to deal with data representing the same category of things such as students' names and ages in your school. The data will need to be sorted, queried or searched, and accessed easily. And, we sometimes may need to update or delete some of the data.

Java provides a simple data structure called an array to meet these requirements. An array supplies a lot of storage space to accommodate our data. The label of each element of the storage space is called the "index." It is an integer number that starts from 0. The data stored in an array can be all integers, characters, or other types of data.

For example:

```
int[] numbers = new int[7]
```

> → defines an integer array numbers with 7 elements in total

```
char[] letters = new char[4]
```

> → defines a character array letters with 4 elements in total

© Ron Dai 2019
R. Dai, *Learn Java with Math*, https://doi.org/10.1007/978-1-4842-5209-3_28

There are different ways to assign or update element values in an array.

- If you have to assign different values to each element, you will need to declare the array with its size and then assign values to each element like shown here:

```
int[]  numbers = new int[5];
numbers[0] = 1;
numbers[1] = 3;
numbers[2] = 2;
numbers[3] = 4;
numbers[4] = 5;
```

or:

```
int[] numbers = new int[ ] { 1, 3, 2, 4, 5 };
```

- If there is a clear pattern of values in the array elements, you may assign the values in the following way:

```
int[]  numbers = new int[7];
for (int i = 0; i < numbers.length; i++) {
        numbers[i] = 2 * i + 1;
}
```

The property of the array, numbers.length, stores the size value of the array numbers.

We can define the size of the array from input during runtime as shown here:

```
int  k = scan.nextInt();
int[] numbers = new int[k];
```

Example

Which of the following choices is the correct syntax for declaring and initializing an array of 8 integers?

(a) `int a[8];`

(b) `[]int a = [8]int;`

(c) `int[8] a = new int[8];`

(d) `int[] a = new int[8];`

(e) `int a[8] = new int[8];`

Answer

(d)

Lab Work

1. Write a line of code to declare and initialize an integer array variable named data with the element values as 7, -1, 13, 24, and 6.

2. Write code that creates an array named odds that stores all odd numbers between -16 and 48 into it using a for loop. Make sure the array has exactly the right capacity to store these odd numbers.

Problems

1. Which of the following choices is the correct syntax for initializing an array of five integers with a list of specific values?

(a) `int a { 14, 88, 27, -3, 2019 };`

(b) `int[] a = new { 14, 88, 27, -3, 2019 } [5];`

(c) `int[5] a = { 14, 88, 27, -3, 2019 };`

(d) `int[] a = { 14, 88, 27, -3, 2019 };`

(e) `int[] a = new int[ ] { 14, 88, 27, -3, 2019 };`

2. What element values do the array numbers have after the following code is executed?

```
int[] numbers = new int[8];
numbers[1] = 4;
numbers[4] = 99;
numbers[7] = 2;
        int x = numbers[1];
numbers[x] = 44;
numbers[numbers[1]] = 11;
```

CHAPTER 29

Common Pitfalls

In this chapter, I want to share several pieces of code that expose common issues in coding practice. Using these examples to diagnose root causes will help improve your understanding. I recommend thinking independently before seeking answers. You may find some hints in the final chapter.

Lab Work

1. Anything wrong here?

```java
String aAsString;
String bAsString;

Scanner user_input = new Scanner(System.in);

System.out.println("a=");
aAsString = user_input.next();
a = Integer.valueOf(aAsString);

System.out.println("b=");
bAsString = user_input.next();
b = Integer.valueOf(bAsString);
```

© Ron Dai 2019
R. Dai, *Learn Java with Math*, https://doi.org/10.1007/978-1-4842-5209-3_29

2. Any errors here?

```
public class TestArray {
        public static void main(String[] args) {
                int[] myArray = new int[] { 11, 12, 13,
                14, 15 };
                System.out.printf("%d\n", myArray[5]);
        }
}
```

3. Understand what the following function is trying to do and think about how to improve it.

```
public static int CountStrings(String[] stringsArray,
String countMe) {
        int occurences = 0;
        if (stringsArray.length == 0) {
                return occurences;        // or, return 0;
        }
        for (int i = 0; i < stringsArray.length; i ++) {
                if (stringsArray[i].toLowerCase().
                contains(countMe.toLowerCase())) {
                        occurences ++;
                }
        }
        return occurences;
}
```

4. Spot the defect:

```
public class Rectangle {
        public int width;
        public int height;
        public int getArea() {
```

```
                    return width*height;
        }
}

public class SomethingIsWrong {
        public static void main(String[] args) {
                Rectangle myRect;
                myRect.width = 40;
                myRect.height = 50;
                System.out.println("myRect's area is "
                + myRect.area());
        }
}
```

5. Spot the defect:

```
Scanner newscanner = new Scanner(System.in);
System.out.print("Please enter today's date (month
day):");
int z = newscanner.nextInt();
int y = news scanner.netInt();
if (z > 12 || y > 31) {
        System.out.println("You have entered an invalid
        number.");
        return;
} else if (y > 31 && z > 12) {
        System.out.println("Both numbers you have
        entered are invalid.");
        return;
}
```

6. Spot the defect:

```
System.out.println("What month were you born in?
(1-12)");
Scanner sc = new Scanner(System.in);
String a = sc.nextLine();
Integer result = Integer.valueOf(a);
int al = result.intValue();
```

7. Spot the defect:

```
if (numToTake >= 2 && numToTake< 3) {
        numToTake = 2;
} else if (numToTake > 2) {
        System.out.println("The number you have entered
        is invalid.");
}
```

CHAPTER 30

Design Considerations

We have learned some fundamental concepts about classes and objects in Java. Now let's look at several examples from the class design perspective.

Practical Case 1

The following is a design of a Rectangle class. It wants to compute a rectangle's area, perimeter, and diagonal, given its width and height values as input parameters.

```java
public class Rectangle {
    private int width;
    private int height;
    private int area;
    private double diagonal;
    private int perimeter;

    public Rectangle (int width, int height) {
        this.width = width;
        this.height = height;
        this.area = width*height;
```

© Ron Dai 2019
R. Dai, *Learn Java with Math*, https://doi.org/10.1007/978-1-4842-5209-3_30

```
            this.diagonal = Math.sqrt(width * width + height
            * height);
            this.perimeter = (width + height) * 2;
    }

    public int getArea() {
            return this.area;
    }
    public double getDiagonal() {
            return this.diagonal;
    }
    public int getPerimeter() {
            return this.perimeter;
    }
}
```

The computations of area, parameter, and diagonal are being done inside the Rectangle constructor, which is executed every time an object of the Rectangle class is initialized. It works if we consistently want to get the values of the area, perimeter, and diagonal of the rectangle. But when we sometimes only want to query the area, perimeter, or diagonal of the rectangle, some part of the computations become excessive. A much better design approach is an "on-demand" implementation as shown here.

```
public class Rectangle {
        private int width;
        private int height;

        public Rectangle (int width, int height) {
                this.width = width;
                this.height = height;
        }
```

```
public int getArea() {
    return this.width * this.height;
}
public double getDiagonal() {
    return Math.sqrt(this.width * this.width
    + this.height * this.height);
}
public int getPerimeter() {
    return (this.width + this.height) * 2;
}
}
```

Practical Case 2

The following example is an implementation of a Game class design. It looks good except for a couple of private field type design choices.

-   The price for goods is usually a small integer plus two decimal places to the right of the decimal point. Neither float types nor double types can accurately represent this form of number used for money calculations because of floating-point inaccuracies. It is recommended to represent the dollar price in cents, so you only need the program to take care of the integer computations. On some occasions, computing money in dollars may be good enough.

– The gameType should not be defined as a true/false
 Boolean value. It should use "String" data type. (Or, we
 may consider using enumeration, if we have a known
 list of fixed names for the gameType.)

```
public class Game {
        private int price;
        private boolean gameType;
        private String platform;

        public Game() { }

        public int getPrice() {
                return this.price;
        }
        public int setPrice(int price) {
                return this.price=price;
        }
        public boolean getGameType() {
                return this.gameType;
        }
        public boolean setGameType(boolean gameType) {
                return this.gameType=gameType;
        }
        public String getPlatform() {
                return this.platform;
        }
        public String setPlatform(String platform) {
                return this.platform=platform;
        }
}
```

Practical Case 3

How do we test a class we have designed in Eclipse?

There are at least two simple approaches. Assume you have designed a class called MyClass. It has one public integer data field - myNumber, and one method to double its integer number value - doubleMe().

Approach A

Both the original class and test code are contained in one Java file as shown here:

```java
public class MyClass {
    // class design part of code
    public int myNumber;
    public MyClass() {      }
    public int doubleMe() {
        return this.myNumber * 2;
    }

    // test part of code
    public static void main(String arg[]) {
        // declare and initialize an object
        MyClass myObject = new MyClass();
        myObject.myNumber = 2019;
        int output = myObject.doubleMe();
        // output the resulting data and validate it
        System.out.println("My result is: " + output);
    }
}
```

Approach B

The following two classes are in separate Java files:

In MyClass.java:

```
public class MyClass {
        public int myNumber;
        public MyClass() {
        }
        public int doubleMe() {
                return this.myNumber * 2;
        }
}
```

In TestMyClass.java:

```
public class TestMyClass {
        public static void main(String arg[]) {
                MyClass myObject = new MyClass();
                myObject.myNumber = 2019;
                int output = myObject.doubleMe();
                System.out.println("My result is: "  + output);
        }
}
```

Practical Case 4

What are the differences between a static and a non-static field or method? And, when do we use static fields and static methods?

In most of the code examples depicted earlier, we used non-static fields and methods (a.k.a. instance fields and instance methods). Both an instance field and an instance method belong to the object instantiated, which means they are not activated until after the object has been created.

However, static fields and static methods belong to the class level. They can be accessed by class name, instead of by any object instantiated from the class. The values stored in static fields and computed by static methods are shared among all objects created from the same class.

The first and most familiar static method to us is "main()" method, if you recall. It can reside in any public class. This method is a unique entry point of any application. It has to be associated with a class. In other words, it doesn't live in any object instance.

In the Demo class example, there is a static field counter that tracks the number of objects created during runtime. There is a non-static field (i.e., instance field) - myNumber that is associated with an individual object instance. The non-static method (i.e., instance method) - getNumber() also belongs to the object created.

```
public class Demo {
        private static int counter;
        public static int getCounter() {
                return counter;
        }

        private int myNumber;
        public int getNumber() {
                return this.myNumber;
        }

        public Demo(int number) {
                this.myNumber = number;
                counter++;
                System.out.println("I am no. " + counter + "
                object so far.");
        }
}
```

The next is a test class to demonstrate how the static field (i.e., counter) and the static method (i.e., Demo.getCounter()) work, in comparison to the non-static field (i.e., myNumber) and the non-static method (i.e., getNumber()).

```
public class TestDemo {
        public static void main(String[] args) {
                Demo demo1 = new Demo(21);
        System.out.println("demo1 myNumber: " + demo1.getNumber());
                System.out.println("object counts: " + Demo.
                getCounter());

                Demo demo2 = new Demo(57);
                System.out.println("demo2 myNumber: " + demo2.
                getNumber());
                System.out.println("object counts: "
                + Demo.getCounter());

                Demo demo3 = new Demo(99);
                System.out.println("demo3 myNumber: " +
                demo3.getNumber());
                System.out.println("object counts: " + Demo.
                getCounter());
        }
}
```

The output from the console is:

```
I am no. 1 object so far.
demo1's myNumber: 21
object counts: 1
I am no. 2 object so far.
```

```
demo2's myNumber: 57
object counts: 2
I am no. 3 object so far.
demo3's myNumber: 99
object counts: 3
```

CHAPTER 31

IOU Computation

IOU means "Intersection Over Union." It is used as a metric in image detection technology. This metric computes a ratio of the overlap area between two rectangles over their union area. For simplicity, the two rectangles are in the same direction, as you will see R1 and R2 in Figure 31-1.

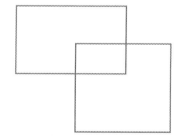

Figure 31-1. *Two rectangles and their overlap*

To figure out this ratio, we need to find out their overlap area named X. If the areas for the two rectangles are R1.area and R2.area, then

$$IOU = X / (R1.area + R2.area - X)$$

We define the location of a rectangle by x_min, y_min, x_max, and y_max. Its four vertices can be represented by the four coordinates: (x_min, y_min), (x_min, y_max), (x_max, y_max), (x_max, y_min), started from the left bottom vertex, going clockwise.

Let's first find out under what circumstances there will be no overlap area between R1 and R2, as shown in Figure 31-2.

© Ron Dai 2019
R. Dai, *Learn Java with Math*, https://doi.org/10.1007/978-1-4842-5209-3_31

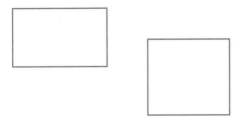

Figure 31-2. *Two rectangles that are apart from each other*

It will be when:

R1.x_max <= R2.x_min,   (1)

or R1.x_min >= R2.x_max, (2)

or R1.y_max <= R2.y_min, (3)

or R1.y_min >= R2.y_max (4)

If one of the conditions from (1) to (4) is valid, the overlap area is 0.

Next, we notice that the overlap area is actually surrounded by four lines, as shown in Figure 31-3.

x = max(R1.x_min, R2.x_min), x = min(R1.x_max, R2.x_max)

y = max(R1.y_min, R2.y_min), y = min(R1.y_max, R2.y_max)

Figure 31-3. *Two rectangles and their overlap areas*

According to the mathematical inferences, we can come up with a coding design solution as shown:

There are two classes, `Rectangle` and `IntersectionOverUnion`.

The `Rectangle` class defines a data model for a rectangle on an x-y coordinate system.

```
public class Rectangle {
        public float x_min;
        public float x_max;
        public float y_min;
        public float y_max;

        public Rectangle(float xmin, float ymin, float xmax,
        float ymax) {
                if (xmin >= xmax || ymin >= ymax) {
throw new IllegalArgumentException("Not a valid rectangle!");
                }
                this.x_min = xmin;
                this.y_min = ymin;
                this.x_max = xmax;
                this.y_max = ymax;
        }

        public float getWidth() {
                return this.x_max - this.x_min;
        }
        public float getHeight() {
                return this.y_max - this.y_min;
        }
}
```

The IntersectionOverUnion class contains the main() method, which drives the execution.

```
public class IntersectionOverUnion {
    public static void main(String[] args) {
        // test case 1
        Rectangle r1 = new Rectangle(3f, 2f, 5f, 7f);
        Rectangle r2 = new Rectangle(4f, 1f, 6f, 8f);
        System.out.println("IOU=" + getIOU(r1, r2));

        // test case 2
        r1 = new Rectangle(3f, 2f, 5f, 7f);
        r2 = new Rectangle(1f, 1f, 6f, 8f);
        System.out.println("IOU=" + getIOU(r1, r2));

        // test case 3
        r1 = new Rectangle(3f, 2f, 5f, 7f);
        r2 = new Rectangle(6f, 1f, 7f, 8f);
        System.out.println("IOU=" + getIOU(r1, r2));
    }

    public static float getIOU(Rectangle r1, Rectangle r2) {
        float areaR1 = r1.getHeight() * r1.getWidth();
        float areaR2 = r2.getHeight() * r2.getWidth();
        float overlapArea = 0f;
        if (r1.x_min >= r2.x_max || r1.x_max <= r2.x_min ||
                r1.y_min >= r2.y_max || r1.y_max <= r2.y_
            min) {
            return 0f;
        }
        overlapArea = computeOverlap(
                        Math.max(r1.x_min, r2.x_min),
                        Math.min(r1.x_max, r2.x_max),
```

```
                              Math.max(r1.y_min, r2.y_min),
                              Math.min(r1.y_max, r2.y_
                              max));
        System.out.println(overlapArea + " / (" + areaR1
                      + " + " + areaR2 + " - " +
                      overlapArea + ")");
        return overlapArea / (areaR1 + areaR2 -
        overlapArea);
    }

    private static float computeOverlap(
                              float x1,
                              float x2,
                              float y1,
                              float y2) {
        float w = x2 - x1;
        if (w < 0) w = -w;
        float h = y2 - y1;
        if (h < 0) h = -h;
        return w * h;

    }
}
```

We are not done yet. We need to always think about how to improve our class design and optimize code. In the Rectangle class, there are getWidth() and getHeight() methods. What if we add a method called getArea() to the Rectangle class?

The Rectangle class is updated as:

```
public class Rectangle {
        public float x_min;
        public float x_max;
        public float y_min;
        public float y_max;
```

```java
    public Rectangle(float xmin, float ymin, float xmax,
    float ymax) {
            if (xmin >= xmax || ymin >= ymax) {
throw new IllegalArgumentException("Not a valid rectangle!");
            }
            this.x_min = xmin;
            this.y_min = ymin;
            this.x_max = xmax;
            this.y_max = ymax;
    }

    public float getWidth() {
            return this.x_max - this.x_min;
    }
    public float getHeight() {
            return this.y_max - this.y_min;
    }
    public float getArea() {
            return this.getWidth() * this.getHeight();
    }
}
```

And the rest of the code will look like:

```java
import java.lang.Math;
public class IntersectionOverUnion {
    public static void main(String[] args) {
            // test case 1
            Rectangle r1 = new Rectangle(3f, 2f, 5f, 7f);
            Rectangle r2 = new Rectangle(4f, 1f, 6f, 8f);
            System.out.println("IOU=" + getIOU(r1, r2));

            // test case 2
```

```java
        r1 = new Rectangle(3f, 2f, 5f, 7f);
        r2 = new Rectangle(1f, 1f, 6f, 8f);
        System.out.println("IOU=" + getIOU(r1, r2));

        // test case 3
        r1 = new Rectangle(3f, 2f, 5f, 7f);
        r2 = new Rectangle(6f, 1f, 7f, 8f);
        System.out.println("IOU=" + getIOU(r1, r2));
    }

    public static float getIOU(Rectangle r1, Rectangle r2) {
        float areaR1 = r1.getArea();
        float areaR2 = r2.getArea();
        float overlapArea = 0f;
        if (r1.x_min >= r2.x_max || r1.x_max <= r2.x_min ||
                r1.y_min >= r2.y_max || r1.y_max <= r2.y_
                min) {
            return 0f;
        }
        overlapArea = computeOverlap(
                        Math.max(r1.x_min, r2.x_min),
                        Math.min(r1.x_max, r2.x_max),
                        Math.max(r1.y_min, r2.y_min),
                        Math.min(r1.y_max, r2.y_
                        max));
        System.out.println(overlapArea + " / (" + areaR1
                + " + " + areaR2 + " - " +
                overlapArea + ")");
        return overlapArea / (areaR1 + areaR2 -
        overlapArea);
    }
```

```
private static float computeOverlap(
                            float x1,
                            float x2,
                            float y1,
                            float y2) {
        float w = x2 - x1;
        if (w < 0) w = -w;
        float h = y2 - y1;
        if (h < 0) h = -h;
        return w * h;
    }
}
```

The computation of area is now encapsulated inside the Rectangle class. This change itself is not big, but we should get used to making small changes at a time when we are still able to incrementally improve our program design.

CHAPTER 32

Projects

I want to recommend a list of hands-on projects for you to practice independently. Working through these projects will definitely help you deepen your understanding of the basic Java programming concepts described in this book.

Project A
Step 1

Write a class called Rectangle that represents a rectangular two-dimensional region. The constructor creates a new rectangle whose top-left corner is specified by the given coordinates and with the given width and height.

```
public Rectangle(int x, int y, int width, int height)
```

Your Rectangle objects should have the following methods:

- public int getHeight() - Returns this rectangle's height.

- public int getWidth() - Returns this rectangle's width.

- public int getX() - Returns this rectangle's x-coordinate.

© Ron Dai 2019
R. Dai, *Learn Java with Math*, https://doi.org/10.1007/978-1-4842-5209-3_32

- public int getY() - Returns this rectangle's y-coordinate.

- public String toString() - Returns a string representation of this rectangle, such as:

  "Rectangle[x=1,y=2,width=3,height=4]"

Step 2

Add the following accessor methods to your Rectangle class from the previous exercises:

```
public boolean contains(int x, int y)
public boolean contains(Point p)
```

The Point class has been defined as shown:

```
public class Point {
        private int x;
        private int y;

        public Point(int x, int y) {
                this.x = x;
                this.y = y;
        }
        public int getX() {
                return x;
        }
        public int getY() {
                return y;
        }
}
```

The two contains() methods return a Boolean state of whether the given Point or coordinates lie inside the bounds of this Rectangle or not. For example, a rectangle with [x=2, y=5, width=8, height=10] will return true for any point from (2, 5) through (10, 15) inclusive, which means the edges are included.

Project B

Design a program to find the number of days between the current day and the user's birthday, given four input values.

The program prompts for the user's birthday. The prompt lists the range of values from which to choose. Notice that the range of days printed is based upon the number of days in the month the user typed. The program prints the absolute day of the year for the birthday. January 1st is absolute day #1 and December 31st is absolute day #365. Last, the program prints the number of days until the user's next birthday. Different messages appear if the birthday is today or tomorrow. The following are four runs of your program and their expected output (user input data is right after the '?' mark):

```
Please enter your birthday:
What is the month (1-12)? 11
What is the day (1-30)? 6
11/6 is day #310 of 365.
```

Your next birthday is in 105 days, counted from today.

Project C

The game rule is this: you start with 21 sticks, and two players take turns either taking one or two sticks. The player who takes the last stick loses. Can you design a program to simulate one of the two players in the game? One player is a user and the other player is the computer.

Project D

Write a method named hasVowel() that returns whether a string has included any vowel (a single-letter string containing a, e, i, o, or u, case-insensitively).

Project E

Write a method named gcd() that accepts two integers as parameters and returns the greatest common divisor (GCD) of the two numbers. The GCD of two integers a and b is the largest integer that is a factor of both a and b. The GCD of any number and 1 is 1, and the GCD of any number and 0 is the number.

One efficient way to compute the GCD of two numbers is to use Euclid's algorithm, which states the following:

GCD(A, B) = GCD(B, A % B)

GCD(A, 0) = Absolute value of A

For example:

- gcd(24, 84) returns 12

- gcd(105, 45) returns 15

- gcd(0, 8) returns 8

Project F

Write a method named toBinary() that accepts an integer as a parameter and returns a string of that number's representation in binary. For example, the call of toBinary(42) should return "101010".

Project G

Use the four numbers on the following cards to create a math expression that equals 24. Each card can be used only once. Treat ace as a number "1". You may use +, -, *, /, (and) in the math expression. Please find all possible answers.

CHAPTER 33

Java Intermediate Solutions

For your reference, in this chapter I'll provide you with answer hints to some of the problems in the earlier chapters. For example, "For 16." means "Hints for problems in Chapter 16."

For 16. Pythagorean Triples

1. Instead of using "c," we may check whether $(a^2 + b^2)$ is a perfect square number, which is taking a square root of it and validating if it is an integer value.

2. Use the example code and check whether the resulting value of $(a^2 + b^2)$ matches the form of "4n + 1".

For 17. Strong Typed Programming

```
public boolean isCollinear(Point p) {
        if (p.getX() == p1.getX() && p1.getX() ==
        p2.getX()) {
                return true;
        }
```

```
        if (this.getSlope(p) == this.getSlope()) {
                return true;
        }
        return false;
    }

public double getSlope(Point p) {
    if (this.p1.x == this.p.x) {
            throw new
IllegalStateException("Denominator cannot be 0");
    }
    return (double)(this.p.y - this.p1.y) / (this.p.x -
    this.p1.x);
}
```

For 18. Conditional Statements

1.   It is rewritten as shown here.

```
        if (num < 10 && num > 0) {
                System.out.println("It's a one-digit
                number");
        }
        else if (num < 100) {
                System.out.println("It's a two-digit
                number");
        }
        else if (num < 1000) {
                System.out.println("It's a three-digit
                number");
        }
```

```
else if (num < 10000) {
        System.out.println("It's a four-digit
        number");
}
else {
        System.out.println("The number is not
        between 1 & 9999");
}
```

2. A simplified version is shown here.

```
if (a == 0) {
        if (b == 0) {...}
        else {...}
} else {
        if (b != 0) {...}
}
```

For 19. Switch Statement

```
switch(color) {
    case 'R':
            System.out.println("The color is red");
            break;
    case 'G':
            System.out.println("The color is green");
            break;
    case 'B':
            System.out.println("The color is black");
            break;
```

```
case 'C':
default:
        System.out.println("Some other color");
        break;
}
```

For 21. Counting

1. Define x as the number of children and $(2200 - x)$
    is the number of adults, then $1.5 * x + 4 * (2200 - x)$
    $= 5,050$. Iterate $x = 0$ up to 2200 to find a solution for x.
    And it is obvious that there is no more than one
    solution.

2. Define x as the number of correct answers and
    $(10 - x)$ as the number of incorrect answers, then
    $5 * x - 2 (10 - x) = 29$. Iterate x from 0 up to 10 to find
    a possible solution for x.

3. Iterate a positive integer from 0 to 2001 and check its
    divisibility with 3, 4, and 5.

4. Iterate every three-digit integer number, from 100
    up to 999, and check its digits.

5. Use a recursive method (referring to the example)
    to repeatedly pick a plant five times from the three
    types of plants (defining three types as A, B, C).
    And then remove duplicates from the combinations.
    For example: {A, A, B, B, C} is a duplicate of {A, B, A,
    B, C}.

For 23. Exploratory Experimentation of Pi

Utilize the following formula with integer number "r" and approximate the value of "e."

$$e = 1 + \frac{1}{1!} + \frac{1}{2!} + \frac{1}{3!} + \ldots \frac{1}{r!}$$

For 24. Classes in Object-Oriented Programming

1. a)

2. b)

3.

```
NumberHolder nh = new NumberHolder();
Nh.anInt = 5;
Nh.aFloat = 3.2;
System.out.printIn("anInt=" + Nh.anInt + "; aFloat=" +
Nh.aFloat);
```

4. (A), (D)

5. (B)

For 26. Inheritance – Code Reuse

1. (c)

2. (b), (d), (e), (f)

For 27. Encapsulation and Polymorphism

1.

```java
public interface GeometricObject {
public abstract double getPerimeter();
    public abstract double getArea();
}
```

2.

```java
public class Circle implements GeometricObject {
private final double PI = 3.14159;
protected double radius;
public Circle(double radius) {
        this.radius = radius;
}

// Implement methods defined in the interface
@Override
public double getPerimeter() {
return 2 * PI * this.radius;
}

@Override
public double getArea() {
        return PI * this.radius * this.radius;
}
}
```

For 28. Array – a Simple and Efficient Data Structure

1. (d)

2. $\{0, 4, 0, 0, 11, 0, 0, 2\}$

For 29. Common Pitfalls

1. If you want to get an integer value, why not take an integer input at the beginning?

    This is a corrected version. It is significantly simplified.

    ```
    Scanner user_input = new Scanner(System.in);
    System.out.println("a=");
    int a = user_input.nextInt();
    System.out.println("b=");
    int b = user_input.nextInt();
    ```

2. Does myArray[3] equal "13"?

    Pay attention to the definition of the index of an array element.

3. Is it necessary to check stringsArray.length = 0? And, is it a good approach to do countMe.toLowerCase() inside the for-loop?

This is a recommended version:

```java
public static int CountStrings(String[] stringsArray,
String countMe) {
        int occurences = 0;
        String keyword = countMe.toLowerCase();
        for (int i = 0; i < stringsArray.length; i ++) {
                if (stringsArray[i].toLowerCase().
                contains(keyword)) {
                        occurences ++;
                }
        }
        return occurences;
}
```

4. Has the myRect ever been initialized?

    There is an important line to update in the main()
    method as shown here:

```java
public class SomethingIsWrong {
        public static void main(String[] args) {
                Rectangle myRect = new Rectangle();
                myRect.width = 40;
                myRect.height = 50;
                System.out.println("myRect's area is
                " + myRect.area());
        }
}
```

5. Since the variable temp has been assigned with the
    value of the first element in array1, do we need to
    iterate from i=0 inside the for-loop?

    The simple fix is to change from for (int i = 0;
    ... to for (int = 1; ... in the original function
    as shown.

```java
public static int getMaxLength(ArrayList<String>
array1) {
        if(array1.isEmpty()) {
             return 0;
        }
        else {
             String temp= array1.get(0);
             for (int i = 1; i < array1.size(); i++) {
                  if (array1.get(i).length() >
                  temp.length() ) {
                       temp= array1.get(i);
                  }
             }
             return temp.length();
        }
}
```

6. Check the if/else clause.

    The scope of "y > 31 && z > 12" is already covered
    by the scope of "z > 12 || y > 31". Therefore,
    the "else if (...)" part in the original code is
    meaningless.

7. Review the actual usage of the `Scanner`.

    Due to the same reason stated in 1, the code can be
    corrected as shown:

    ```
    System.out.println("What month were you born in?
    (1-12)");
    Scanner sc = new Scanner(System.in);
    int al = sc.nextInt();
    ```

8. Check the `if/else` clause

    The scope of `numToTake > 2` has included the scope
    of `numToTake >= 2 && numToTake < 3`. The `if` and
    `else if` conditional clauses need to be rewritten.

Index

A

Abstraction, 171, 182
Algorithms
    creation, real-world
        objects, 39, 40
    swap values, 40–41
Array, 223
    character, 185
    data types, 185
    defined, 185
    element values, 186
    index, 185
    size, 186

B

Basic projects, 85–87

C

class instantiation, 29
Class variables/instance
    variables, 33
Coding mistakes, 73, 74
Coding structure, 81
Coin flip game, 93, 94, 96

Collatz conjecture
    defining, 5
    program, 5, 6
Collinearity, 107
Conditional operators, 64–67
Conditional statements, 218, 219
    bigger number identification, 109
    example, 111, 114
    if clauses, 112
    if/else if structure, 110
    if/else structure, 109, 110
    nested if/else structure, 110
    quadrants, 114, 115
    tree-like structure, 111
contains() method, 213
convertToBaseN() method, 144
countBase10Numbers() method, 143
Counting, 220
    countNumbers2(), 136
    for-loop, 137, 139, 140
    isDistinct(...), 135
    single loop, 131
    switch statement, 145
    tables, 132
    tickets, 131
Curly braces, 82

© Ron Dai 2019
R. Dai, *Learn Java with Math*, https://doi.org/10.1007/978-1-4842-5209-3

Printed in the United States
By Bookmasters